T0140806

Josefa Oberem

Examining auditory selective attention: From dichotic towards realistic environments

Logos Verlag Berlin GmbH

λογος

Aachener Beiträge zur Akustik

Editors:
Prof. Dr. rer. nat. Michael Vorländer
Prof. Dr.-Ing. Janina Fels
Institute of Technical Acoustics
RWTH Aachen University
52056 Aachen
www.akustik.rwth-aachen.de

Bibliographic information published by the Deutsche Nationalbibliothek

The Deutsche Nationalbibliothek lists this publication in the Deutsche Nationalbibliografie; detailed bibliographic data are available in the Internet at http://dnb.d-nb.de .

D 82 (Diss. RWTH Aachen University, 2020)

ISBN 978-3-8325-5101-8
ISSN 2512-6008
Vol. 33

Logos Verlag Berlin GmbH
Comeniushof, Gubener Str. 47,
D-10243 Berlin
Tel.: +49 (0)30 / 42 85 10 90
Fax: +49 (0)30 / 42 85 10 92
http://www.logos-verlag.de

Examining auditory selective attention: From dichotic towards realistic environments

Von der Fakultät für Elektrotechnik und Informationstechnik der
Rheinischen-Westfälischen Technischen Hochschule Aachen
zur Erlangung des akademischen Grades einer

DOKTORIN DER INGENIEURWISSENSCHAFTEN

genehmigte Dissertation

vorgelegt von

Josefa Oberem

M.Sc.

aus Bonn, Deutschland

Berichter:

Univ.-Prof. Dr.-Ing. Janina Fels

Univ.-Prof. Dr. phil. Iring Koch

Tag der mündlichen Prüfung: 24.Januar 2020

Diese Dissertation ist auf den Internetseiten der Hochschulbibliothek online verfügbar.

Meinen Eltern gewidmet

Abstract

The aim of the present thesis is to examine the cognitive control mechanisms underlying auditory selective attention by considering the influence of variables that increase the complexity of an auditory scene. Therefore, technical aspects such as dynamic binaural hearing, room acoustics and head movements as well as those that influence the efficiency of cognitive processing are taken into account. Step-by-step the well-established dichotic-listening paradigm is extended into a "realistic" spatial listening paradigm.

Conducted empirical surveys are based on a paradigm examining the intentional switching of auditory selective attention. Spoken phrases are simultaneously presented by two speakers to participants from two of eight azimuthal positions. The stimuli are phrases that consist of a single digit (1 to 9, excluding 5), in some experiments followed by either the German direction "UP" or "DOWN". A visual cue indicates the target's spatial position, prior to auditory stimulus onset. Afterwards, participants are asked to identify whether the target number is arithmetically smaller or greater than five and to categorize the direction.

Human performance measure differences in reaction times and error rates between the repetition of the target's spatial position and the related switch (i.e. switch costs) describe the loss of efficiency associated with redirecting attention from one target's location to another. To examine whether the irrelevant auditory information is decoded, interference in the processing of task-relevant and task-irrelevant information is created in the paradigm.

Using the binaural-listening paradigm, the ability to intentionally switch auditory selective attention is tested when applying different methods of spatial reproduction. Essential differences between real sources, an individual and a non-individual binaural synthesis reproduced with headphones as well as a binaural synthesis based on Cross-Talk Cancellation are found. This indicates how the loss of individual information reduces the ability to inhibit irrelevant information. As a step towards multi-talker scenarios in realistic environments participants are tested in differently reverberating environments. Switch costs are highly affected by reverberation and the inhibition is also impaired by to be unattended information. Age-related effects are also found when applying the binaural-listening paradigm, indicating difficulties for elderly to suppress processing the distractor's speech.

Contents

1 Introduction

Communication in noisy reverberant environments is an immense challenge for our auditory attention. Referred to as the "cocktail-party effect", it has been in the interest of research since Cherry [26] reported his initial study asking participants to selectively listen to one ear while ignoring the speech from a distracting speaker in the other ear. Using dichotic-listening paradigms, many different facets of auditory attention have been analyzed in the last decades (among others [20, 169, 32, 140, 22]).

Recently, Koch and colleagues [74] applied dichotic listening to examine intentional switching of auditory selective attention. The paradigm is based on the combination of dichotic listening [26] with the methodology of task cueing [107]. Koch and colleagues' auditory task-switching paradigm differs from other studies on attention switches (for example [81, 150, 165]). These studies deal with involuntary attention switches, meaning that the attention switches are not instructed but occurred spontaneously. In contrast, Koch and colleagues explicitly emphasize the examination of endogenous, voluntary attention switches.
In the present paradigm attention switches are cued in advance and referred to the target's gender or the target's location, indicating that the target's location/gender is switched or repeated between subsequent trials. To be more precise, a switch of the target's location means that the target is positioned to the left side in the preceding trial and in the following trial the target is positioned to the right side. Further studies [72, 73, 85, 89, 86, 88, 87, 162, 163, 161, 164] that use the introduced dichotic-listening paradigm report about their main finding on a cued switch of the relevant target which resulted in a worse performance than a cued repetition of the relevant target's speaker gender.
To examine whether the irrelevant auditory information is encoded, an interference in the processing of task-relevant and task-irrelevant information is created in the paradigm. The participants' task is to categorize the spoken digit (1 to 9, excluding 5) presented by the target speaker into categories of smaller or greater than five. To respond to the task the associated response button has to be pressed. The two simultaneously presented stimuli of one trial are either

congruent or incongruent. To be more precise, for congruent trials digits are either both smaller than five or both greater than five and therefore call for the same response. In incongruent trials, one digit is smaller and one is greater than five and therefore call for different responses. Participants' performance measures are smaller in congruent trials than in incongruent trials which is numerously confirmed [74, 72, 73, 85, 89, 86, 88, 87, 162, 163, 161, 164]. The "congruency effect" [71] suggests the lack of inhibition and therefore a processing of irrelevant information [134].

The dichotic-listening paradigm on intentional switching of auditory selective attention has several advantages: it is technically very easy to handle and convenient, it uses experimentally well-controlled stimuli, and it is capable of very precise performance measures. However, to completely understand the cognitive control mechanisms underlying auditory selective attention in realistic environments utilizing dichotic listening is not sufficient. A dichotic presentation is a highly artificial situation compared to natural listening. A realistic "cocktail-party" scenario includes a number of additional cues that are associated with binaural hearing.

To study the binaural effects in the intentional switching of auditory selective attention, the dichotic-listening paradigm is gradually extended towards a binaural-listening paradigm representing complex dynamic acoustic scenes in the present thesis [129, 44, 134, 136, 45].

In order to realize the extension of the paradigm towards a realistic scene various technical methods and tools need to be applied. As the listening paradigm is step-wise broadened towards realistic scenes the technical methods and tools are assessed with respect to the collected empirical results.
The advantages and shortcomings of individual compared to non-individual head-related transfer functions (HRTFs) have been in the focus of research for several decades. Usually differences and similarities are found using localization tasks (among others [160, 25, 180, 115]). Furthermore, studies on plausibility and authenticity were applied to evaluate the needed accuracy of HRTF measurements [38, 39, 41, 42, 126, 131, 186, 119, 82, 156, 94, 19]. However, a simple localization task or comparisons of differently plausible stimuli lack in representing a listening task in complex environments. Applying different binaural reproduction methods to the paradigm on the intentional switching of auditory selective attention is the approach of this thesis to gain a deeper insight.
To create reverberant and dynamic binaural scenarios further software tools are necessary. In the present thesis, *RAVEN* (Room Acoustics for Virtual ENvi-

ronments) [159] and Virtual Acoustics (*VA*) [65, 179] are used, which have been developed at the Institute of Technical Acoustics, RWTH University Aachen. By applying these software tools to the binaural-listening paradigm on auditory selective attention benefits and deficiencies are analyzed.

This thesis describes and evaluates the step-by-step development of the binaural-listening paradigm and the conducted application scenarios.
Chapter 4.1 and 4.4 evaluate the general extension from the dichotic-listening paradigm to the binaural-listening paradigm. Different binaural reproduction methods are compared utilizing the newly developed binaural-listening paradigm in chapter 4.2 to examine how they affect the results in experiments involving auditory selective attention. Since an anechoic spatial reproduction of stimuli fails to represent a realistic multi-talker conversation in a noisy environment, reverberant energy is provided to observe whether auditory selective attention is affected in chapter 4.3 and 4.5. Preceding results imply an analysis of head-movements in a dynamic binaural reproduction, which is discussed in chapter 4.6. In chapter 4.7 and 4.8 the binaural-listening paradigm is also applied to older participants to explore age-related effects in auditory selective attention in spatial environments.

2

Fundamentals

This chapter gives a brief introduction into the fundamentals used in this thesis. After defining different methods of auditory reproductions, the theory of binaural hearing and Head-Related Transfer Function (HRTF) are introduced. Furthermore, some relevant background information on terms from experimental psychology are edited.

2.1. Definition of Auditory Reproduction

Stimuli can be presented monaurally or binaurally to a listener. Monaural refers to a presentation relating to only one ear and binaural to both ears. In a binaural reproduction, the stimuli can either be identical, called diotic or different, called dichotic [12].
In experimental psychology, stimuli are often presented monaurally or dichotically in listening experiments via headphones (for example [26, 20, 22, 140, 74]).
A spatial, binaural presentation of stimuli, which is by definition dichotic, since left and right ear's stimulus are not identical, is usually used in technical acoustics (for example [116, 53, 129]). In technical acoustics and also in the present thesis, the terms **binaural** and **dichotic** are used slightly deviating from the formal definition. Using **binaural** it is only referred to the situation where stimuli reach both ears and also include spatial information. The term **dichotic** is used referring to two different stimuli presented separately to the two ears excluding the special case of stimuli containing spatial information [45].

2.2. Fundamentals of Spatial Hearing

2.2.1. Head-Related Coordinate System

To describe the relation between a listener and sound sources, the head-related coordinate system is introduced, depicted in figure 2.1. The center of the coordinate system is placed in the middle of the head between the upper edge of the entrances of the ear canals [11]. The interaural axis passes through all

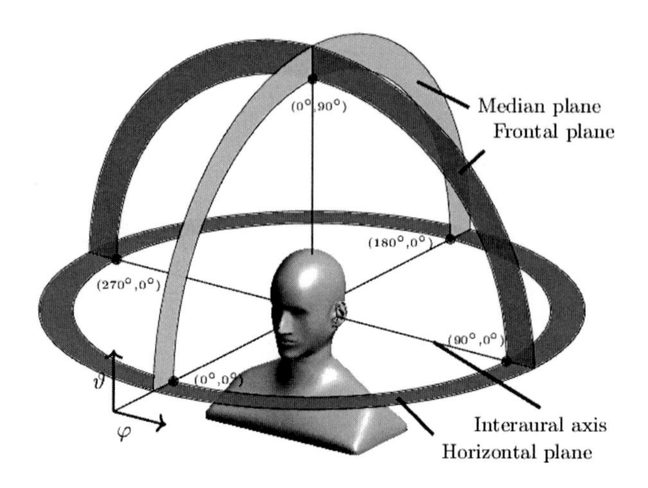

Figure 2.1.: Head-related coordinate system definitions [148].

those three points and spans the horizontal plane with the front-back connection (compare figure 2.1, colored in red). The frontal plane divides the forehead and face from the back of the head along the interaural axis (compare figure 2.1, colored in blue). The median plane cuts the head along the front-back axis into two symmetrical halves (compare figure 2.1, colored in green).

2.2.2. Binaural Hearing

The ability to hear binaurally makes it possible to localize sound sources. Subtle differences in intensity, spectral, and timing cues enable a listener to aurally find a position in space.

Interaural Time Difference (ITD)

The arrival-time of a sound wave is in most cases not identical for left and right ear, due to different path lengths from the source to the ears. This arrival-time difference is called Interaural Time Difference (ITD). The maximal ITD ($\sim 690\,\mu$s [120]) is given, when a sound source is positioned on the interaural axis, directly facing one ear. The sound wave has to travel all around the head to arrive at the opposite ear. In contrast, sound waves from a source positioned in the median plane arrive simultaneous and therefore the Interaural Time Difference dissolves ($ITD = 0$). Azimuthal localization is therefore mainly based on the ITD cue [13, 120].

Interaural Level Difference (ILD)

The sound wave is always disturbed by reflections and diffraction of the listener's head and torso depending on its direction of arrival. This causes an attenuation especially at the averted ear, which is called the Interaural Level Differences (ILD). The level difference or intensity difference is frequency dependent. Low frequencies ($< 500\,\mathrm{Hz}$) are practically not constrained by the body, higher frequencies ($> 1.5\,\mathrm{kHz}$) are highly affected by reflections, diffraction and head shadowing. Frequencies greater than $3\,\mathrm{kHz}$ are even influenced by the shape of the pinna. The ILD cue provides information for localization in horizontal and vertical angle [13, 120].

Duplex Theory

Already in 1907, Lord Rayleigh [95] reported his so called "duplex theory", indicating how the physical cue of ILD is most useful at high frequencies, while the cue of ITD is most useful at low frequencies. Nowadays, the definition of the duplex theory is loosened and it is recognized that the ITD is also important in the higher frequency range [30] . The duplex theory is not strictly accurate for complex sounds and the discrimination of front-back-reversals can also not be explained by the duplex theory.

Cone of Confusion

Front-back-reversals often occur when ITDs and ILDs are identical for two or more positions. It is also referred to "cones of confusion" to describe positions of indistinguishable interaural differences. This holds true for all positions placed on the surface of circular conical slices centered around the interaural axis. The median plane is a special case of cone of confusions where ITD and ILD equal zero. Ambiguities related to the cone of confusion may be resolved by monaural cues and small head movements [120].

Monaural Cues

The outer ear and especially the shape of the pinna form a direction-selective filter. Complex concave folds of the pinna reflect and scatter the incoming sound wave depending on the the direction of the acoustic source. These amplifications and attenuations of particular frequencies are called monaural cues, which are mainly important for vertical sound localization [24].

Sounding Movements

The advantage from small spontaneous head movements may assist resolving ambiguities on cones of confusion. This observation originally is reported by van Soest in 1929 [173] and later systematically described by Wallach [177]. It was often shown how prominent localization errors such as front-back uncertainties can be solved by sounding movements [14, 97, 183].
For stimuli with a duration of less than 600 – 800 ms [66, 143, 147] there is no gain for dynamic cues since it takes a minimum time of 200 ms to initiate head movements [11]. These sounding movements are mainly rotations performed in horizontal plane [168] resulting in ITD and ILD changes. Blauert [11] observed, how at least 95 % of these movements are smaller than 1 °.

Head-Related-Transfer-Function (HRTF)

The head-related transfer function (HRTF) describes the filtering process of incoming sound waves due to diffraction and interference at the body, head and outer ear. A free-field HRTF is defined by the sound pressure measured at the eardrum or at the entrance of the ear canal divided by the sound pressure measured with a microphone in the center of the head-related coordinate system with the head absent.
The HRTF $\underline{H}(f)$ is the Fourier transform of the Head-Related Impulse Response (HRIR) $h(t)$. In figure 2.2 an example of an HRTF and the belonging HRIR is shown, as well as the procedure of transformation of time domain and frequency domain of a linear and time invariant system.

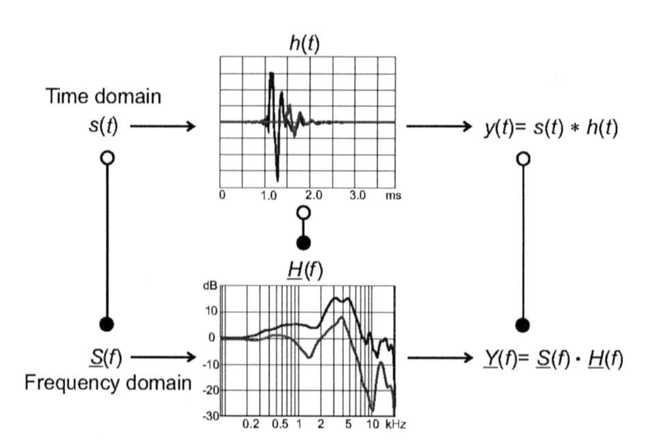

Figure 2.2.: Linear time-invariant system of HRTF and HRIR [37].

2.2.3. Binaural Synthesis

To localize a sound source in space humans merely process the differences between the signals at the two ears, described above. To create a virtual sound source which is localized in a certain direction by the listener, these interaural differences and the monaural coloration need to be integrated into the presented monaural stimulus. Due to channel separation a common and convenient way of reproduction is to convolve (non-)individual HRTFs with the desired stimulus and to play back over headphones.

Artificial Heads

Often HRTFs are measured with artificial heads (compare figure 3.11) instead of measuring the HRTFs of real participants due to the lack of time or for practical reasons. Findings in localization tasks applying individual and non-individual HRTFs show that differences are significant [116, 109]. Throughout the present thesis an artificial head produced at the Institute of Technical Acoustics, RWTH Aachen University, with a simple torso and a detailed ear geometry [158, 111] is used, when refering to non-individual HRTFs.

Headphone-Transfer Function (HpTF)

When reproducing binaural synthesis via headphones, headphone characteristics and the transfer path from membrane to the eardrum are naturally included into the presented binaural synthesis. Colorations based on the frequency response of the headphone, that mainly occur in higher frequencies, are not desirable. The HRTF already contains the transfer path along the pinna to the eardrum. It is not preferable to add these amplifications and attenuations from reflections of the outer ear. By measuring a headphone transfer function (HpTF) and convolving the inverse with the binaural synthesis this challenge is resolved. This is also described as a headphone equalization [102, 113].

2.3. Psychological Background

2.3.1. Historical Beginning of Studying Auditory Selective Attention

Examining auditory selective attention has a long tradition in experimental psychology. Referred to as the "cocktail-party effect", this has been in the focus of research since Cherry [26] reported his first dichotic-listening study where participants are asked to selectively listen to one ear while ignoring the speech from a distracting speaker in the other ear. The speakers are always assigned to one of the ears. The participant's task is to directly shadow and therefore to

repeat the relevant speech out loudly. During the reproduction, the stimulus of the distracting speaker is partly switched to a different language or the speaker's gender is changed. After the shadowing task, participants are asked about the to be unattended speech of the opposite ear. They are mostly able to report whether the gender of the distracting speaker has changed and therefore sensory characteristics of the irrelevant information are noticed. However, characteristics that require perceptual processing, as for example the language of the irrelevant speech, cannot be processed. Furthermore, participants are not able to report the content of the speech for the most parts [62, 140, 87].
An important research question is therefore, how much of the to be unattended information is actually processed and how much could be ignored or is filtered out [61]. Three different theories are delineated, subsequently.

Early Selection or Filter Theory

Based on Cherry's findings [26] and his own auditory attention examination using digit pairs, Broadbent [20] proposed the early filter theory. The theory posits that stimuli are filtered at an early stage of processing based on basic physical features, such as color, pitch or incoming direction. Consequently, the theory implies that the relevant and irrelevant information is processed ear-wise, in serial order. A selective filter provides the transmission of irrelevant-ear information into a buffer store and the relevant-ear information towards perceptual processing. Hence, the unattended ear information is not processed beyond the in higher perceptual levels [81]. Broadbent substantiated his theory using a split-span technique, where participants listened to different lists of digits presented dichotically. It was observed that listeners reported the digits of one ear first and than switched to the other ear [20]. Shortcomings of the theory were detected in other empirical studies (among others [169]), which is why further theories are postulated.

Attenuation Theory

In contrast to the early filter theory, Treisman [169] as well as Moray [121] found attention switches to the to be unattended ear after the participant's name is included in the irrelevant speech, meaning some of the irrelevant information must be processed perceptually. Treisman proposed an alternative theory, called attenuation theory. This theory supports the early filter theory, however, it assumes an attenuation of the irrelevant speech up to the level of perceptual processing, rather than filtering out. To gain conscious awareness of the to be unattended information, the information needs to surpass a threshold. Depending on the semantic features of the information the threshold is differently easy to pass. For example if the information includes the listener's name the threshold is comparably low and the information can be attended.

Late Selection Theory

A third theory was proposed in 1963 by Deutsch and Deutsch [32]. It is assumed that a selective filter restrains the irrelevant information, however, at a later stage in information processing compared to Broadbent's early filter theory. In more detail, it is argued that information is selected after processing for meaning and therefore all information is attended up to the point where semantic encoding and analysis is performed. Afterwards only the most important stimuli are selected for further processing.

2.3.2. Control of Processing Irrelevant Information

To control and examine whether and how much of the irrelevant auditory information is nevertheless encoded, an interference in the processing of task-relevant and task-irrelevant information is created in the paradigm used in the present thesis [74, 45] (compare chapter 3.1.1 and 3.2.2). The impairment of the performance is measured by the congruency effect [71]. Koch and colleagues [74] found in several studies how congruent and incongruent stimulus reproduction differed in performance [72, 73, 85, 89, 86, 88, 87, 162, 163, 161, 164]. For congruent stimulus reproduction relevant and irrelevant auditory information result in the same response categories of task requirement, whereas for incongruent stimulus reproduction relevant and irrelevant auditory information result in different response category of task requirement. The congruency effect can be interpreted as an implicit performance measure of attending to task-irrelevant information or in other words disobeying the task instructions [74, 45].

2.3.3. Maintaining and Switching Attention

The ability to enhance the processing of certain stimuli while suppressing the information from another concurrent stimuli is called selective attention. An interesting research question is how well a certain stimuli can be attended. Substantial research effort has been invested in the observation of maintenance of attention on one sound source and hence, the prevention of attention switches towards another sound source [184, 150]. These attention switches are not instructed but occurred spontaneously and involuntary [81, 150, 165, 22].
In contrast, Koch and colleagues [74] explicitly examined endogenous, voluntary attention switches and thus a paradigm to explore the intentional switching of auditory selective attention is introduced.

Fundamentals of the Present Paradigm

The present paradigm (compare chapter 3.1) utilizes the intentional switching of auditory selective attention combined with the methodology of task cueing [107, 67, 118, 174]. Cued attention switches referred to the target's gender or the target's location [73, 85]. A target's location switch implies that the target's location switched between trials, whereas in the preceding trial the target is for example on the left side and in the subsequent trial the target is on the right side (compare also chapter 3.2.1). The main finding is that a cued switch of the relevant target's location or gender resulted in a worse performance than in cued repetitions of the relevant target's location or gender [72, 74, 85, 88]. The performance costs are observed in two performance measures: reaction time and accuracy. This is different than in several other studies on auditory selective attention [165, 22], where speech perception and comprehension are measured directly in terms of the accuracy of (verbal) report. The present paradigm offers the possibility to observe and interpret reaction times, since auditory attention is assessed by requiring participants to categorize the target's stimulus as quickly as possible while ignoring the simultaneously presented distracting speaker's stimulus [45].

Switch Costs

Switch costs describe the loss of efficiency associated with redirecting attention from one target's location to another. Therefore, performance measures provide worse results (i.e. longer reaction times and higher error rates) when the target's location is switched between trials compared to a repetition of the target's location. Switch costs describe the numerical difference in performance measures between switches and repetitions. The switch costs point to cognitive interference in information processing when the selection criterion needs to be intentionally adjusted and provide an explicit measure of how well instructions to switch attention can be followed.

Exogenous and Endogenous Cues

To cue the auditory attention-switch a visual cue is presented to the participant (compare chapter 3.1). Throughout this thesis endogenous cues are used. However, in discussed literature (for an overview [70]) exogenous cues are also applied and confined to the present work.
Exogenous cues are often used in detection tasks and lead to an automatic (i.e., bottom up) target selection. Therefore, exogenous cues initiate an involuntary movement of attention. An example for such a cue in a spatial listening experi-

ment would be a small flashing/active LED lamp mounted on top of the active loudspeaker.
In contrast, endogenous cues instruct a participant in a task to direct attention to a particular location. Attention to "actively" select (i.e., top down) the target stimulus is needed in order to perform the task. Therefore, endogenous cues initiate a voluntary movement of attention. An example for such a cue in a spatial listening experiment would be a visual symbolic cue at a centered screen indicating the position of the active loudspeaker.

2.3.4. Age-related Effects in Auditory Attention Switching

Aging humans largely experience a decline in cognitive capacity to some degree, that becomes apparent through forgetfulness, a decreased ability to maintain focus as well as a decreased problem solving capacity. Demanding situations as for example communication in noisy environments evoke difficulties for senior citizens. It has been proven that elderly adults often have difficulties to follow a conversation in a multi-talker situation [57, 172]. These communication challenges may possibly arise from an age-related hearing loss, however, a constraint in cognitive processing may be another reason [145, 144]. The theory of general slowing and the inhibitory deficit theory try to provide answers for some effects of the decline in cognitive capacity.

General Slowing

Generally the theory of general slowing states and explains the poorer cognitive performance of older adults on a variety of cognitive tasks. The general hypothesis states that increased age in adulthood is associated with a decrease of the speed of cognitive processing. Several investigations [153, 154, 175] observe the general slowing of information-processing speed. A way to address to this effect, when analyzing reaction time data of different age groups, is comparing the logarithmic-transformed scores of young and old participants [78, 104].

Inhibitory Deficit Theory

An important mechanism behind the concept of cognitive control is the inhibition. To be able to focus attention the irrelevant information needs to be inhibited. When talking about the ability to switch attention and to ignore distracting speakers in elderly, the inhibitory-deficit theory becomes important [54, 96]. The theory implies an age-related deficit in inhibitory processes, which are important for multi-talker situations in complex environments. Inhibition functions are the access, the deletion and the restrain of information that decline with age according to Braver and Barch [17] as well as others [55].

3

Experimental Setups

This chapter defines and specifies the paradigm, variables, data filters, stimuli, laboratories, measuring procedures, reproduction techniques and roomacoustics modeling software used for the experiments on intentional switching of auditory selective attention.

3.1. Paradigm

The paradigm was firstly introduced by Koch and colleagues [74] and developed to analyze the intentional switching in auditory selective attention using dichotic listening. Stepwise the paradigm is transformed into a binaural-listening paradigm representing more realistic scenes than a dichotic-listening scenario. Experiments described in this thesis are based on one of the following three versions of the paradigm: dichotic-listening paradigm, binaural-listening paradigm and extended binaural-listening paradigm.

3.1.1. Dichotic-Listening Paradigm

The dichotic-listening paradigm consists of two simultaneously presented stimuli. These stimuli are delivered by two speakers of opposite sex and are presented each at one ear.
"One speaker acts as the target and the other acts as the distractor. The participant is asked to focus on the target speaker and ignore the distracting speaker. To distinguish between target and distractor, the target speaker's direction is cued in advance."[129] Hence, a visual cue (letter, compare figure 3.4 (a)) highlighting the target's position is shown on a monitor (15 inch screen, 1.8 m distance in anechoic chamber and 0.7 m distance in hearing booth). The letter L indicates that the participant is asked to focus his/her attention to the left speaker, accordingly the letter R for the right speaker.

The speech material comprises all digits except zero and five (speech material: 1, 2, 3, 4, 6, 7, 8, 9). Every speaker presents one of those eight digits. Spoken digits are never identical within one trial, but can be both smaller than 5, both greater than 5 or one greater and one smaller than 5. The listener's task is to categorize the target's speech into smaller or greater than five. The two stimulus categories are mapped to two response buttons, held in hands, to be pressed by the left and right thumb (compare figure 3.2 (a)). The push button held in the left hand is to be pressed in case the relevant number is smaller than 5 and right response key should be pressed for numbers greater than 5, consistent with the mental number line [127, 40, 129].

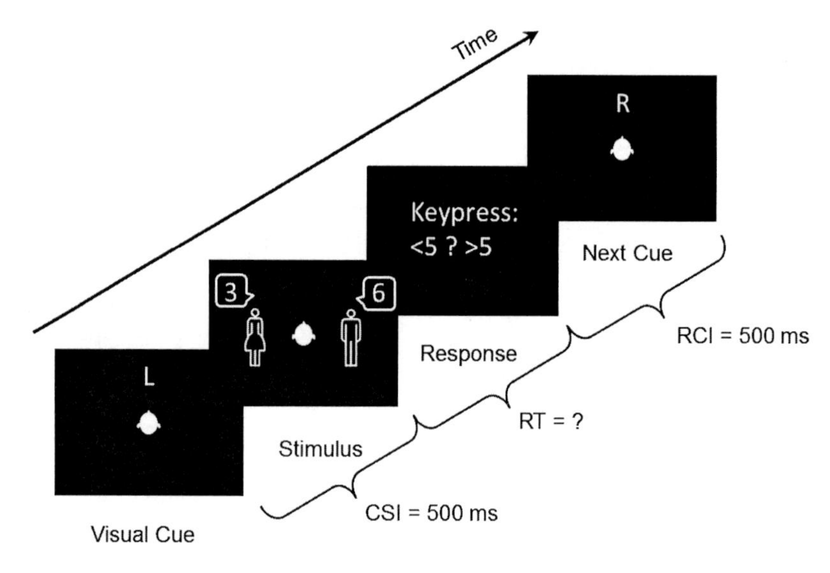

Figure 3.1.: Dichotic-listening paradigm: Procedure of a trial with a visual cue indicating the target direction (L vs. R), a cue-stimulus-interval (CSI) of 500 ms, the synchronous presentation of the stimuli, reaction time between onset of stimulus and the response of the participant, and the response-cue-interval (RCI) of 500 ms.

Figure 3.1 shows the procedure of a trial. Each trial starts with a visual cue presented on the monitor in front of the participant. After a cue-stimulus interval (CSI) of 500 ms, the two acoustic stimuli (target and distractor) are simultaneously presented. The visual cue remains on the screen until the participant responds to the acoustic target. The interval between response and next cue

(RCI) is also set to 500 ms. In case of an error, visual feedback ("Fehler!", German for "error") is displayed for 500 ms, delaying the onset of the next cue.
Dependent variables of the experiment are always measured reaction times (time between onset of stimulus and button press) as well as error rates.

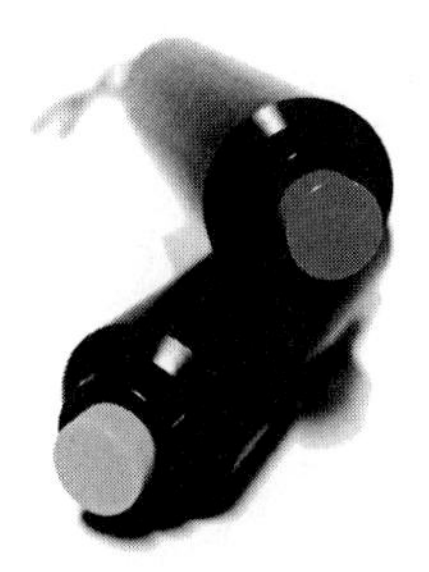

(a) Push buttons for response input.

(b) Controller for response input.

Figure 3.2.: Devices for response input: (a) Push button held in hand to be pressed with right and left thumbs for response input in dichotic-listening and binaural-listening paradigm. (b) Controller used for response input with extended binaural-listening paradigm. Four buttons on front side to be pressed by right and left index and middle finger.

3.1.2. Binaural-Listening Paradigm

The dichotic-listening paradigm is extended into a binaural-listening paradigm, firstly introduced by Oberem and colleagues [129].
The main procedure of a trial of the binaural-listening paradigm is kept identical to the procedure of the dichotic-listening paradigm, but visual cue and speaker positions are changed. There are eight possible positions for target and distracting speaker (front, front-right, right, back-right, back, back-left, left, front-left). Target-speaker and distracting speaker are located in two different directions. Due to the enlargement of possible positions, the visual cue needs to be adjusted. The visual cue consists of a sketch of all directions indicating the target direction with a filled dot (compare figure 3.4 (b)).

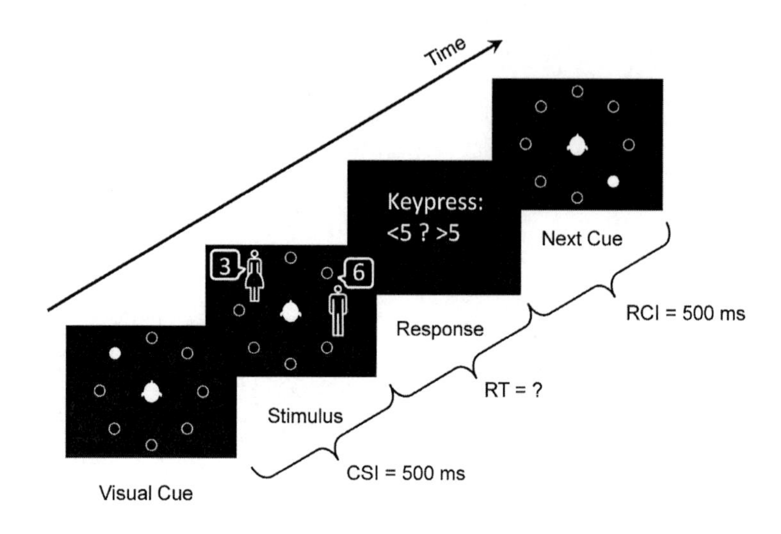

Figure 3.3.: Binaural-listening paradigm: Procedure of a trial with a visual cue indicating the target direction, a cue-stimulus-interval (CSI) of 500 ms, the synchronous presentation of the stimuli, reaction time between onset of stimulus and the response of the participant, and the response-cue-interval (RCI) of 500 ms [129].

(a) Dichotic-listening paradigm.

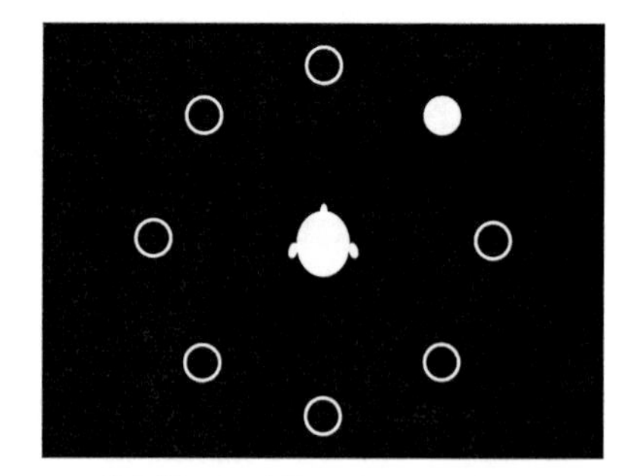

(b) Binaural-listening paradigm.

Figure 3.4.: Visual cue: (a) in dichotic-listening paradigm, indicating the target's position to the right. (b) in binaural-listening paradigm, indicating the target's position in front-right [136, 129, 134].

3.1.3. Extended Binaural-Listening Paradigm

To analyze auditory selective attention in realistic environments, the setup and stimuli are modified stepwise. When using reverberating stimuli it became clear that the binaural-listening paradigm is limited (compare chapter 4.3, [128]. Short and monosyllabic stimuli (730 ms, compare chapter 3.4.1) are not or only neglectably influenced by reverberation. Stimuli and response options are successfully extended to ensure the possibility to analyze the intentional switching of auditory selective attention in realistic environments (compare chapter 4.4, [43, 44]).

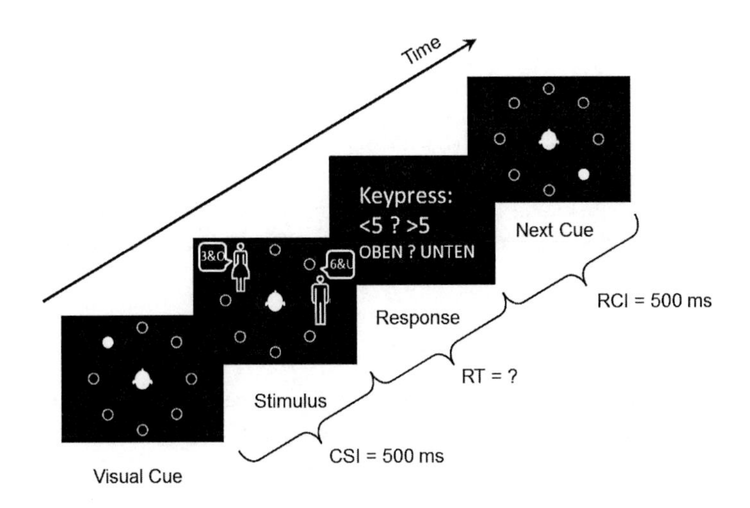

Figure 3.5.: Extended binaural-listening paradigm: Procedure of a trial with a visual cue indicating the target direction, a cue-stimulus-interval (CSI) of 500 ms, the synchronous presentation of the stimuli, reaction time between onset of stimulus and the response of the participant, and the response-cue-interval (RCI) of 500 ms.

In the extended binaural-listening paradigm, "stimuli of both speakers consist of a single spoken digit (1-9, excluding 5) which is followed by a German disyllabic direction word for "UP" or "DOWN" (e.g. the combined stimulus could be "Eight Up" or "One Down") [125]. The participants' task is a two-step process. Concerning the digit the participant has to categorize the relevant digit presented by the target speaker as smaller or larger than five. Secondly, the direction word has to be cognitively processed to press the corresponding response button [(compare figure 3.2 (b))]. There are four response possibilities given in a quadratic arrangement to be pressed by index fingers and middle fingers of both hands (clockwise: “> 5+up” to be pressed by right index finger, “> 5+down” to be pressed by right middle finger, “< 5+down” to be pressed by left middle finger, “< 5+up” to be pressed by left index finger). Therefore the categories of smaller and greater five are mapped to the left hand buttons and the right hand buttons at the front side of a controller. Furthermore, the direction word presented by the target gives information whether the index finger (in case the direction word is “UP”) or the middle finger (in case the direction word is “DOWN”)has to be pressed”[136].

3.2. Independent Variables

Independent variables vary between experiments. Auditory attention switch and congruency are used in every experiment. The spatial position of the target and the spatial angle between target and distractor are variables that are created in the course of the extension into a binaural-listening paradigm. Further independent variables are described in the belonging chapter.

3.2.1. Auditory Attention Switch

Auditory attention switch (AS) refers to the target’s spatial position in two consecutive trials. The variable has two different levels (switch vs. repetition). The target’s spatial position can either be repeated from one trial to another (e.g. front - front) or switched between trials (e.g. left - back). The distractor’s position is switched between all trials (compare figure 3.6).
The corresponding differences between switch-trials and repetition-trials in reaction times and error rates are also called switch costs [71].

3.2.2. Congruency

“Congruency (C) refers to the stimuli of target and distractor within one trial. The variable has two different levels (congruent vs. incongruent). The two stimuli can be congruent, which is the case when both number words are smaller than five or both greater than five (e.g. 2 and 4, 6 and 9), or they can be incongruent, which is the case when one number word is smaller and one is greater than five (e.g. 1 and 7, 8 and 3) ”[134] (compare figure 3.7).

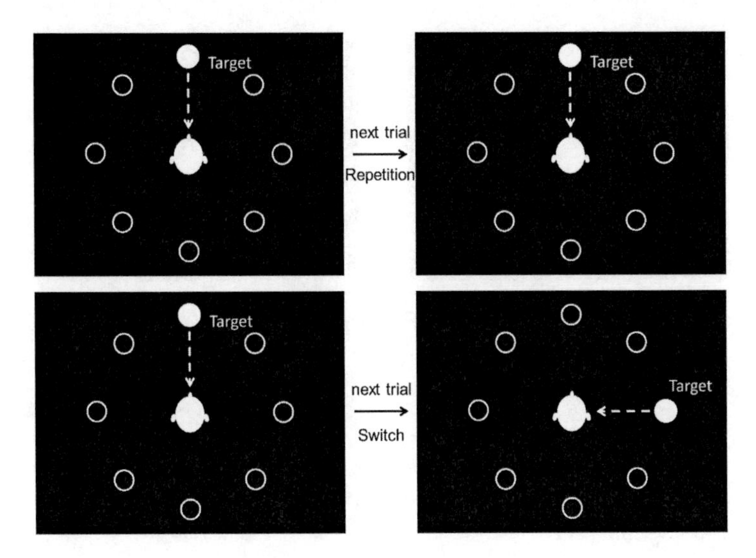

Figure 3.6.: Auditory attention switch refers to the target's spatial position in two consecutive trials. The variable has two different levels (switch vs. repetition).

When the extended binaural-listening paradigm (compare chapter 3.1.3) is applied the congruency needs to be redefined. The variable has still two different levels (congruent vs. incongruent). A trial is considered congruent when target's digit and distractor's digit belong to the same category and the direction word is identical. In case the digits belong to different categories and/or the direction word is not identical the trial is considered as incongruent.

3.2.3. Spatial Position of Target

The effect of the target's position (TPOS) is studied in the binaural-listening paradigm. There are eight possible positions for the target (compare figure 3.4 (b)). "These eight positions are sorted into three different categories (compare figure 3.8). The categories are designed with respect to the planes of the head-related coordinate system [11] as well as the results of the analysis of the first study using the binaural-listening paradigm [128]. The first category included all positions on the median plane (front; back) and was therefore called "median plane". The second category described all positions placed on the inter-aural axis and therefore on the frontal plane (left; right); it is later called "frontal plane".

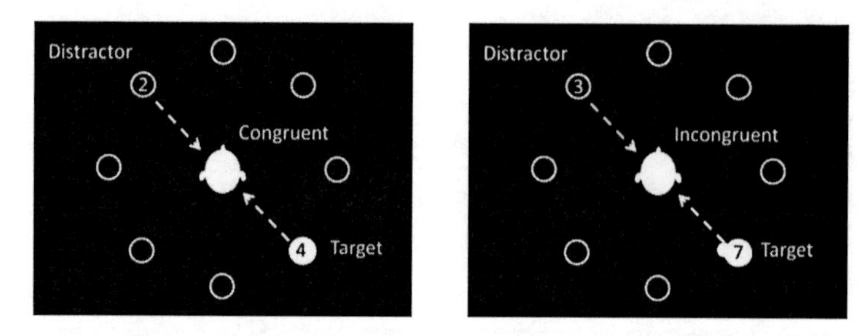

Figure 3.7.: Congruency refers to the stimuli of target and distractor within one trial. The variable has two different levels (congruent vs. incongruent).

The third class, named "diagonal plane", included all other possible spatial positions which were located in 45 ° from the defined planes of the head-related coordinate system"[134].

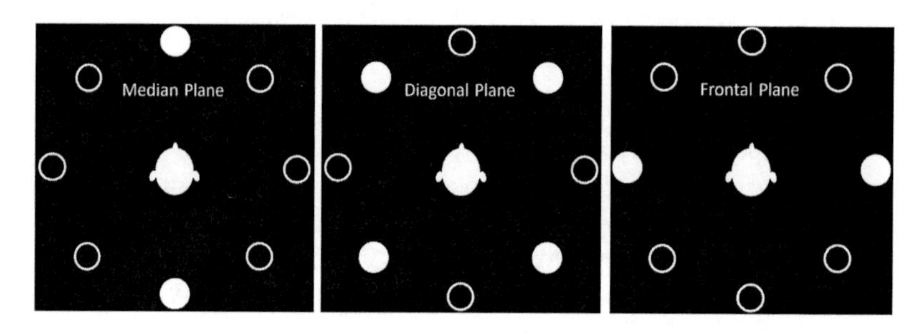

Figure 3.8.: Possible positions of the target speaker categorized in three planes: median plane, diagonal plane and frontal plane [136, 134].

3.2.4. Spatial Angle between Target and Distractor

To analyze the spatial relation of the speaker the variable of angle between target's and distractor's spatial location is introduced. This variable has four levels (compare figure 3.9). Target and distractor can be directly neighbored and therefore separated by 45 °. They can be across from each other (180 °) or perpendicular (90 °). The fourth possibility of the spatial arrangement for target and distractor comprises 135 °.

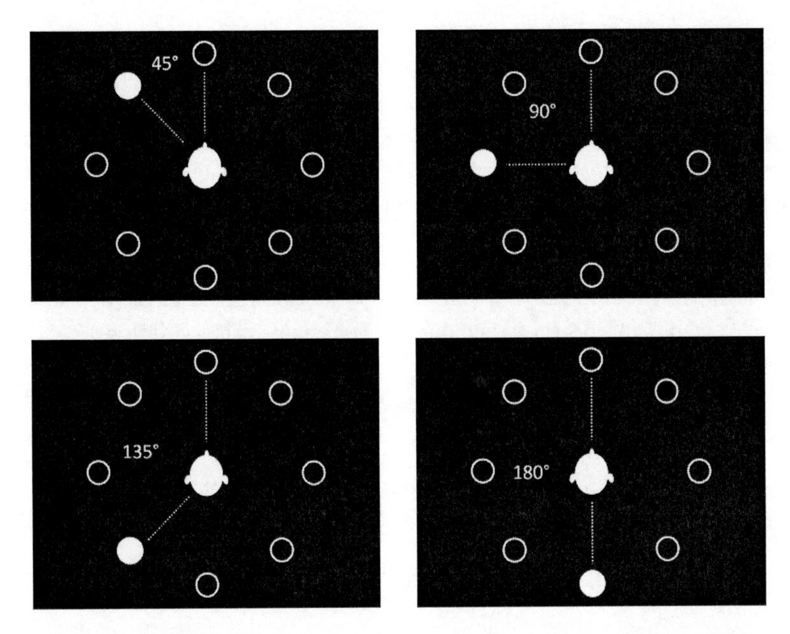

Figure 3.9.: Possible angles between target's and distractor's spatial location: 45 °, 90 °, 135 °, 180 °. Exemplary for the distractor's position being in front. [136]

3.3. Data selection and statistics

For the analysis of reaction times (RT) and error rates (ER), the training sequences are always removed from the data. The first trial in every block and every trial with a reaction time exceeding ±3 standard deviations from the individual's mean reaction time are also excluded from the analysis. In experiments performed in the anechoic chamber using a static reproduction all trials in which head movements ($> \pm 1$ cm in translation and $> \pm 2$ ° in rotation) are detected by the tracker, are also deleted. Additionally, for the analysis of reaction times, every trial with an error and the following trial are eliminated, since these trials can not validly be defined as switch or repetition trials.
Fundamentals of statistical analysis are not introduced in this thesis. An overview on statistics can be found by Bortz and colleagues [16]. However, tests and corrections that are used in the analyses are named in the following for the sake of completeness.

Using the statistic software *SPSS* by *IBM* repeated measures ANOVAs (Analysis of variance) are calculated for the conducted experiments. To test the normality of data distributions a Kolmogorov-Smirnov test ($p > .05$) is used. For independent variables with more than two levels the assumption of sphericity can be violated. In case Mauchly's test indicates the violation, the Huynh-Feldt correction is applied in the ANOVA. For significant main effects with more than two levels as well as interactions post-hoc tests are computed using the Bonferroni-adjustment. For the sake of clarity, non-significant interactions are not listed in the result sections of the experiments, but can be found in tables listed in the appendix A. All collected data of this thesis is published in a supplementary data set combined with a technical report [124].

3.4. Stimulus Material

All stimuli used for the experiments presented in this thesis are recorded in the anechoic chamber (compare chapter 3.5.1).

3.4.1. Binaural-Listening Paradigm

"Speech material [...][is] recorded under anechoic conditions with two male and two female native German speakers. The used hardware, studio microphone TLM170 by *Neumann* and sound card Hammerfall DSP Multiface by *RME*, [...] [allows] recordings with a frequency range from 70 Hz to 20 kHz. The stimuli consist of single spoken digits (1-9, excluding 5). With a time stretching algorithm that maintains the original frequencies of the recording [64], stimuli [...] [are] shortened or extended to 730 ms (max. modification of length: 20%). Therefore, stimuli start and end synchronously when presented at the same time. The loudness of the recorded stimuli [...] [is] adjusted according to DIN 45631 [48]"[129].

3.4.2. Extended Binaural-Listening Paradigm

For the extended paradigm new speech material is recorded under anechoic conditions with two male and two female professional, native German speakers. "The used hardware, a large diaphragm condensor microphone TLM170 by *Neumann* and Zoom H6 Handy Recorder (both: cardioid directivity pattern), [...] [allows] recordings with a frequency range from 70 Hz to 20 kHz. The stimuli consist, as in the original paradigm, of single spoken digits (1-9, excluding 5) which [...] [are] followed by a German disyllabic direction word ("UP", in German "OBEN" and "DOWN", in German "UNTEN"). All stimuli (digits and direction words separately) [...] [are] shortened or stretched to 600 ms (max. modification of length: 32%) with the same time stretching algorithm [64]. The total length of the stimulus [...] [is] therefore 1200 ms. The loudness of the recorded stimuli [...] [is] adjusted according to DIN 45631 [48]"[136]. This speech material is published in a supplementary data set combined with a technical report [125].

3.5. Laboratory

Early experiments are conducted in the fully anechoic chamber and experiments based on the extended binaural-listening paradigm take place in the hearing booth.

3.5.1. Fully Anechoic Chamber

As a laboratory for listening tests "a fully anechoic chamber ($l \times w \times h = 9.2 \times 6.2 \times 5.0$ m^3) with a lower boundary frequency limit of 200 Hz [...] [is] used. The participants [...] [are] asked to sit inside a frame of eight loudspeakers (compare figure 3.10 (a)), which [...] [are] equally distributed over azimuth (every 45°), whereas the distance between the participant and the loudspeakers [...] [is] kept constant at 1.8 m. The chair [...] [is] provided with a backrest, armrests, and an adjustable head rest. An electromagnetic tracker (Polhemus Patriot) [...] [is] used during the HRTF measurements and the listening test to control and supervise the movements of the participant's head. A monitor (15 inch screen) to present written instructions and show the visual cue [...] [is] installed in a distance of 1.8 m distance. [...] To take the focus from vision to audition, lights [...] [are] turned off during the listening test [120, 11]"[129].

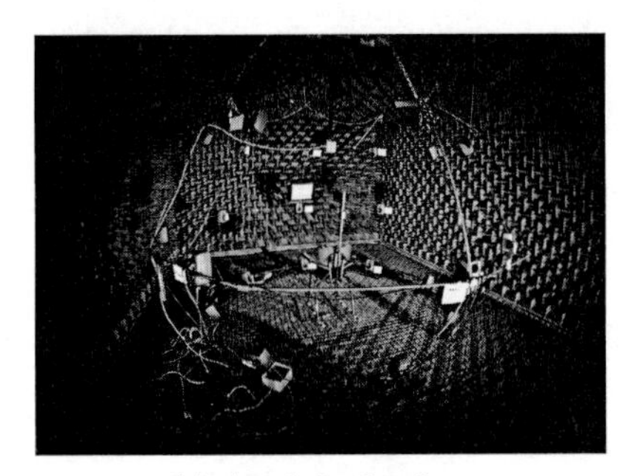

(a) Anechoic chamber.

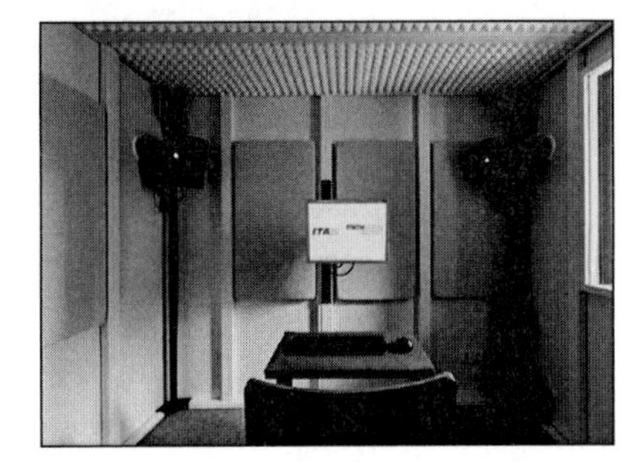

(b) Hearing booth.

Figure 3.10.: Laboratories: (a) Fully anechoic chamber with loudspeaker setup and monitor in front [129]. (b) Hearing Booth for listening tests with monitor in front.

3.5.2. Hearing Booth

To ensure a quiet environment during the test, listening tests also take place in a hearing booth ($l \times w \times h = 2.3 \times 2.3 \times 1.98$ m^3) (compare figure 3.10 (b)). A monitor (15 inch screen) to present written instructions and show the visual cue is installed in a distance of 0.7 m. Again lights are turned off during the listening test.

(a) Headphones with optical markers.

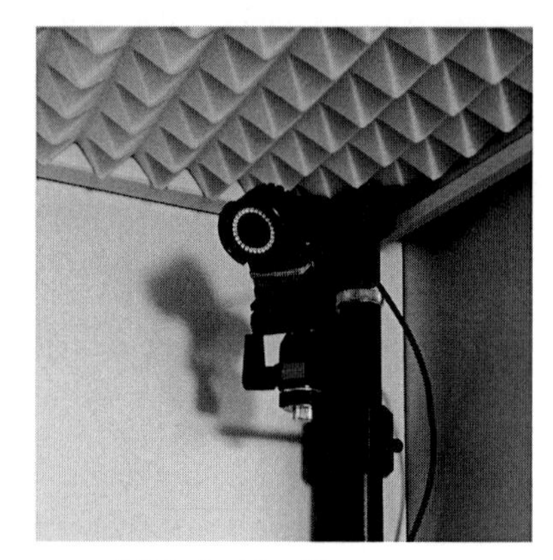

(b) Optical tracking camera.

Figure 3.11.: Tracking devices: (a) Artificial head wearing headphones *Sennheiser* HD600 equiped with retro reflective optical markers by *OptiTrack*. (b) Infrared tracking cameras by *OptiTrack* mounted in the hearing booth to monitor the participants' head-movements.

An optical tracking system (*OptiTrack* by *NaturalPoint Inc.*, [123]) is used to monitor head movements during the experiment. The system used four infrared (wavelength: 850 nm) cameras mounted in the room close to the ceiling (compare figure 3.11b)). The system operates with a sampling frequency of 120 Hz and is capable of monitoring several tracking bodies simultaneously. To monitor the participant's head movements retro reflective optical markers are fastened centrally on the headphones (compare figure 3.11a)). This point is monitored throughout the experiment. As this tracking body is not located in the center of the head, the movement of this tracking body does not correspond in all axes to the actual head movement. To correct this offset, a head calibration is done individually. For this calibration, the position of the two ears is marked at the beginning of the test with two additional tracking bodies. From these two

positions and the position on top of the head, the center point of the head can be approximated more closely.
Three-dimensional position data of the optical markers are registered with the optical sensors, using the triangulation method. The marker positions are displayed with *Motive* which is a software designed to supervise and control motion capture systems for tracking applications.

3.6. Measurements

Depending on the experimental setup HRTFs and/or HpTFs are measured in different settings.

3.6.1. Positioning of Microphones

Different types of microphones used to measure HRTFs within the ear canal as well as the most adequate and applicable position in or around the ear have been investigated by several researchers [51, 106, 110]. Probe microphones were used by Wightman and Kistler [182] as well as Bronkhorst [21] among others due to size and signal to noise ratios, whereas in recent time measurements are more commonly made using miniature microphones placed at the entrance of the blocked ear canal [51, 176]. In 1995, Møller et al. [114] measured HRTFs with an open auditory canal, but reported better results when HRTFs were measured with a blocked ear canal. However, the application and positioning of miniature microphones with silicon Open-Domes (compare figure 3.12 (b), figure 3.13 (b)) is very simple, precise and little time consuming when HRTFs are frequently measured.
Oberem and colleagues [38, 39, 41, 42, 126, 131] compared different recording methods (open meatus vs. blocked meatus) regarding the perception of the spatial sound reproduction. It is found that when using an adequate headphone equalization and a binaural synthesis, the condition of the ear canal and the recording technique do not yield to different findings. The comfortable and little time consuming measuring method using Open-Domes however, is recommended for HRTF and HpTF measurements in terms of plausibility.
In further studies, closed-domes are also used which turned out to be even more practical in precise fastening at the entrance of the ear canal due to the design of two nested domes (compare figure 3.12 (c)).

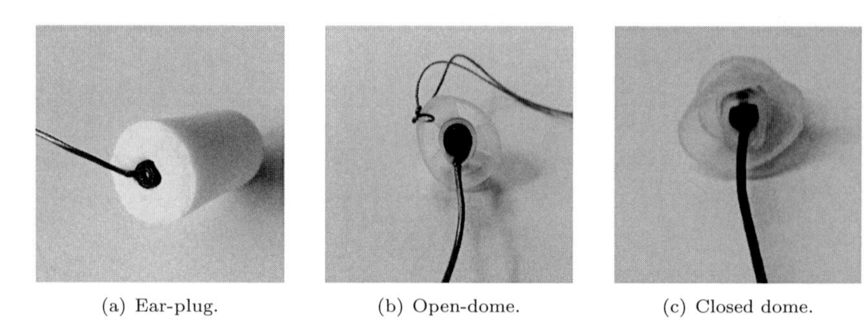

(a) Ear-plug. (b) Open-dome. (c) Closed dome.

Figure 3.12.: Miniature microphones (KE3 by Sennheiser) fixed in (a) ear plug, (b) open dome and (c) closed dome to fasten the microphone flush with the entrance of the ear canal.

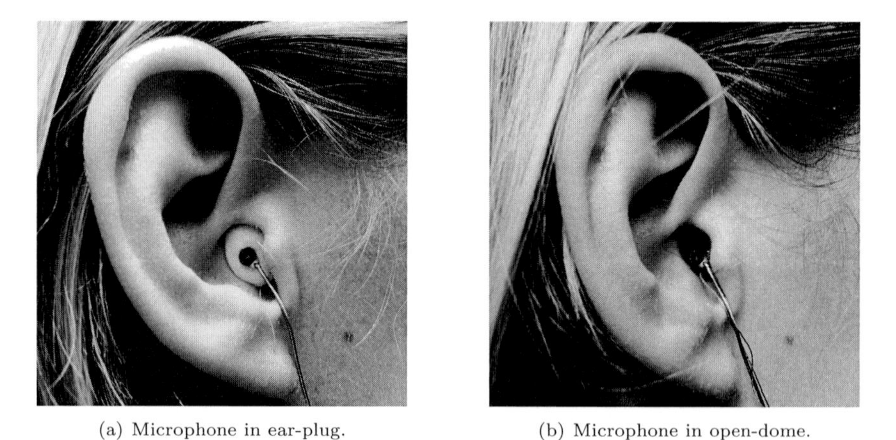

(a) Microphone in ear-plug. (b) Microphone in open-dome.

Figure 3.13.: Miniature microphones (KE3 by Sennheiser) fixed in (a) ear plug and (b) open dome positioned the entrance of the ear canal [131].

3.6.2. HRTF Measurements

Measurement with loudspeakers in loudspeaker arrangement

HRTFs (compare chapter 2.2.2) are either measured in a loudspeaker setup or with the HRTF arc. Individual HRTFs are measured in the anechoic chamber in the construction of loudspeakers (compare figure 3.10). "Measurements [...] [run] automatically with the ITA-Toolbox [64] in *MATLAB*. Interleaved exponential sweeps [33, 98] (frequency range: 70 Hz-20 kHz, bit rate: 24 bit, sampling rate: 44.1 kHz, total excitation length: 7.5 s, no averaging) [...] [are] first sent to the sound card, then converted by an D/A-converter of type *Behringer* ADA8000 Ultragain Pro-8 and amplified, and finally played by the loudspeakers in the anechoic chamber. For recording, microphones (KE3 by *Sennheiser*) [...] [are] placed at the entrance of the ear canal with an open-dome, a little silicon carrier, so that the ear canal [...] [stays] partly open or with an ear plug shortened in length to have a closed ear canal. The in-ear recorded signal [...] [goes] through the above-mentioned A/D-converter and the sound card before being post-processed (including time windowing)."[129]

Measurement with the HRTF Arc

HRTF measurements can also take place in a semi-anechoic chamber ($l \times w \times h = 9.2 \times 6.2 \times 5.0\ \mathrm{m}^3$) with a lower boundary frequency limit of 200 Hz.
The measurement system is specifically developed for fast individual HRTF measurements with a small impact on the measurement signal [149]. It uses 64 loudspeakers (1 ") arranged in an incomplete semi-circle from elevation 0 ° to 160 ° with 2.5 ° degree resolution (compare figure 3.14). The participants have two *Sennheiser* KE3 microphones inserted into their ears at the entrance of the blocked ear canals. They are positioned standing at two meters ear height in the center of the measurement arc on a turntable. The measurement signal is an interleaved sweep of all 64 loudspeakers with a frequency range from 500 Hz to 22050 Hz with a time delay between loudspeaker starts of 40 ms and an overall length of 3.38 s.
To reduce unnecessary head movements, the participant's head is positioned against a head rest. The turntable moves the participant in azimuth angle of 2.5 °. The overall measurement duration is about ten minutes.

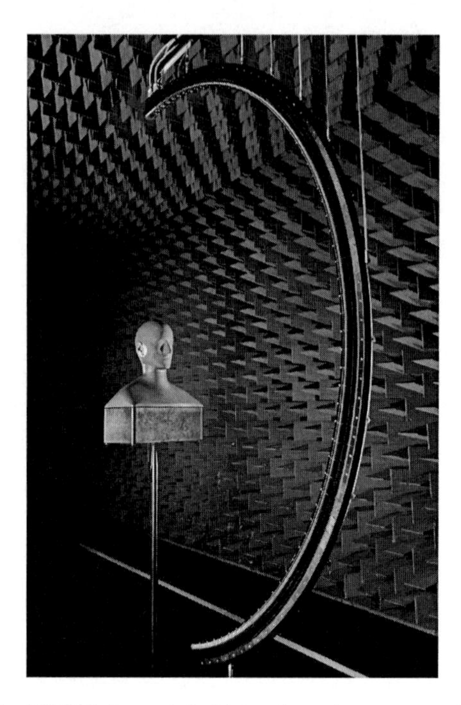

Figure 3.14.: The HRTF Arc with 64 loudspeakers, exemplarily an artificial head is positioned in the center [148].

3.6.3. Individual Headphone Equalization

Headphone-transfer-functions (HpTFs) (compare chapter 2.2.3) are measured to calculate an adequate robust equalization. To this end, miniature microphones (KE3 by *Sennheiser*) in closed domes are positioned in the ear canal to be flush with the entrance. In case HRTF measurements are conducted using open domes or earplugs to fasten the miniature microphones, these are also used for the HpTF measurement. After each of in total eight HpTF measurements, headphones are repositioned on the participants head [103]. To give the best comfort, the repositioning is done by the participant itself. Based on Masiero and Fels [103] the equalization is calculated using the mean of the HpTF measurements. Since phase information is lost at this process, minimum phase is used. Furthermore, notches in the high frequency range are smoothed [103, 131].

3.7. Reproduction Method

Reproduction methods are compared and changed over experiments presented in this thesis.

3.7.1. Dichotic

For the dichotic presentation of stimuli open headphones (Sennheiser HD 600) are used. Although, the characteristics of the used headphones are nearly flat, stimuli for the dichotic reproduction are also convolved with the robust headphone equalization, to assure that no influence by the headphones is given.

3.7.2. Real sources

The scene of competing speakers is reproduced by loudspeakers placed in the anechoic chamber (compare figure 3.10 (a)). “The used loudspeakers [...] [are] *Genelec* two-way active loudspeakers, model 6010A (frequency range: 73 Hz - 21 kHz (-3 dB)), which [...] [are] equalized and fed by the sound card, a Hammerfall DSP Multiface by *RME*.”[129]

3.7.3. Binaural - Static

In most of the described experiments the binaural reproduction of stimuli is static. This means that movements of the participant are not integrated into the binaural scene. In case the participant turns his/her head to one side, the binaural scene turns with the listener, respectively. Therefore, the participant is not free to move within a scene to get for example further localization cues from changing ITD and ILD (compare chapter 2.2.2).

Individual

Individually generated stimuli are presented binaurally via headphones. Therefore, HRTFs are measured individually (compare chapter 3.6.2). Open headphones (*Sennheiser* HD 600) are used for the binaural reproduction and an individual headphone equalization is applied for every participant (compare chapter 3.6.3). The convolution of stimuli, HRTF (filter length: 40 ms) and equalization (filter length: 23 ms) is done off-line with *MATLAB*. Each binaural stimulus is stored as a separate sound file in wave format. The presentation of the binaural stimuli is static [129].

Non-individual

The HRTFs of an artificial head are used to create binaural stimuli. “The dummy head is a mannequin produced at the Institute of Technical Acoustics, RWTH Aachen University, with a simple torso and a detailed ear geometry [158, 111]”[129]. Convolution and processing is identical to the individual binaural stimuli. Stimuli are in most studies presented binaurally via headphones.
In one study the procedure of Cross-Talk Cancellation (CTC) is applied. “First introduced by Atal and Schröder [4], CTC makes it possible to present binaural stimuli via loudspeakers. Detailed information about the theory and procedure can be found in Møller [112], Schmitz [157], and Lentz [91]. A third order CTC filter [...] [is] used and stimuli [...] [are] presented with two loudspeakers in the horizontal plane at $\pm 45^\circ$ (front-left and front-right). Further information about the CTC-filter [...] [are] presented by Majdak et al. [99]. The participant's movements [...] [are] restricted to ± 1 cm in translation and $\pm 2^\circ$ in rotation to ensure that the participant [...] [is] always within the sweet spot [91]”[129].

3.7.4. Binaural - Dynamic

To estimate the relevance of a dynamic reproduction a real-time auralization of the experimental scene is created. In case a scene is reproduced dynamically the listener has the possibility to move freely within the scene. When listening to a static reproduction, however, the listener has a defined position and orientation in the scene. Head movements cause a rotation of the whole scene to the same angle (compare figure 3.15).

Real-time auralization

As participants were free to move their head, different HRTF filters have to be used, depending on the current head position and orientation. To adjust the binaural synthesis synchronized with the head movement, real-time convolution with HRTF filters and headphone equalization is required.
Head movements are monitored and tracked using the *OptiTrack* system (by *NaturalPoint Inc.*, [123]) and the associated software *Motive* (compare chapter 3.5.2).
The real-time auralization was realized with the software Virtual Acoustics (*VA*) which was developed at the Institute of Technical Acoustics, RWTH Aachen University [65, 179]. It allows fast exchange of filters (for latency analysis, see [141, 46]) as a result of head movements and can be controlled by the *OptiTrack* tracking system. A HRTF data set measured with the ITA artificial head [158, 111] with a resolution of $(1^\circ \times 1^\circ)$ is used for the auralization in the present

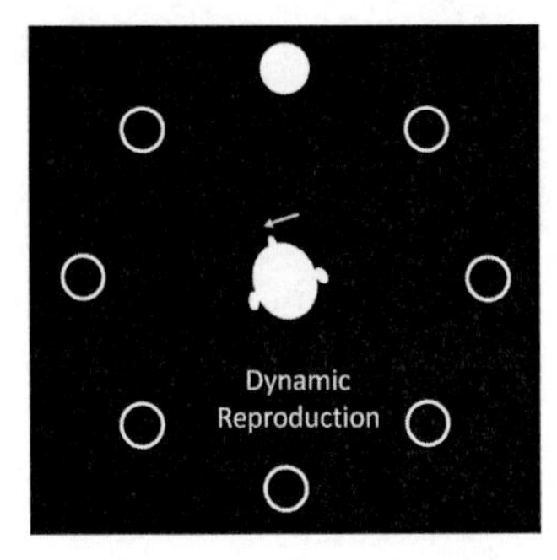

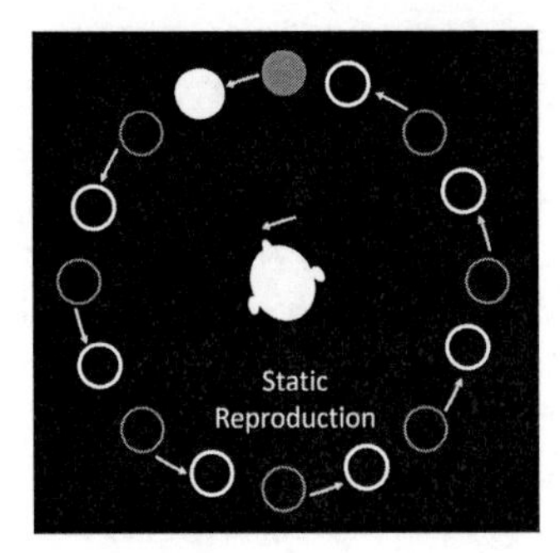

Figure 3.15.: Schematic representation of the consequences of head movements during a dynamic and a static reproduction. In a dynamic scene the listener is able to freely move within the scene. When the listener moves his/her head in a static reproduction the complete acoustical scene turns, accordingly.

thesis. Neighboring HRTFs were selected on a nearest neighbor basis. Therefore, HRTF filters are changed when the participant turns his/her head for more than 0.5 °.

3.8. Roomacoustics

The software *RAVEN* (Room Acoustics for Virtual ENvironments) [159, 2] enables an auralization of reverberating environments. The software is based on the method of deterministic image sources as well as a stochastic ray-tracing algorithm. The auralization software offers a physically nearly accurate auralization of sound propagation in complex environments. It supports spatially distributed and freely movable sound sources and receivers.
For experiments set in reverberating environments presented in this thesis a room model is build and binaural room impulse responses are calculated. "The modeled room has a total volume of 137 m^3 with a quadrangular ground area. All walls are not parallel and have different lengths (front: 6.1 m, right: 7.5 m, back: 6.0 m, left: 7.6 m). The height of the room is set to a constant value of 3 m. The listener is located slightly off the center of the room with a sitting height of 1.3 m (listener's position: l = 2.9 m; w = 3.2 m; h = 1.3 m). The source positions are located in a circular arrangement around the listener, each with a distance of 1.8 m to the listener. The walls are not set to be parallel and the listener is not positioned in the center of the room to prevent unwanted acoustical effects due to nodal points or echos [52, 146, 50].

Absorption coefficients in the room model are changed to achieve three levels of reverberation: anechoic (RT_{60}=0 s), low reverberation (RT_{60}=0.8 s), and high reverberation (RT_{60}=1.75 s). Reverberation times for octave-bands from 63 Hz to 8 kHz are listed in Table 3.1. The absorption coefficients for the three conditions are adjusted according to common building materials for floor, wall and ceiling. Binaural room impulse responses (BRIR) are calculated with *RAVEN* based on the simulated room model as well as HRTFs of an artificial head, measured in an anechoic chamber [159]. The dummy head is a mannequin produced at the Institute of Technical Acoustics, RWTH Aachen University, with a simple torso and a detailed ear geometry [158, 111].
The convolution of stimuli with BRIRs and equalization [(compare chapter 3.6.3)] is done off-line using *MATLAB*. All binaural stimuli are stored separately in wave format" [136].

	Frequency in Hz							
	63	125	250	500	1k	2k	4k	8k
Anechoic	0	0	0	0	0	0	0	0
Low	0.5	0.6	0.8	0.8	0.8	0.8	0.6	0.5
High	2.5	2.2	2.0	1.8	1.7	1.5	1.2	0.8

Table 3.1.: Reverberation times in seconds for the three simulated rooms in octave bands [136].

4

Experiments on auditory selective attention

This chapter describes eight different experiments using the paradigm on intentionally switching auditory selective attention in different settings and versions. Step-wise the dichotic-listening paradigm is transferred into a "realistic-listening" paradigm.

4.1. From Dichotic To Binaural – Experiment I

Parts of this study are presented at the national conference on acoustics DAGA in 2013 [40] and on an international student poster conference in Prague in 2013 [127]. Experimental data can be downloaded from the technical report [124].

The transformation of the dichotic-listening paradigm into a binaural-listening paradigm enabling auditory attention tasks in realistic environments is done step-wise. As the first step the dichotic-listening paradigm is fed with binaural stimuli [127, 40]. More precisely, an experiment on intentional switching of auditory selective attention using the dichotic-listening paradigm is carried out using three different techniques of reproduction: a dichotic reproduction of the stimuli (left and right ear), a reproduction with real sources, in this case loudspeakers located at $\pm 90^\circ$ in the horizontal plane, and a reproduction using individual binaural reproduction via headphones as well as an individual headphone equalization.

To the author's knowledge, there is no investigation comparing dichotical and binaural reproduction methods in an auditory attention task. Several authors already compared the results of monaural and spatial reproduction in different experiments.
Drullman and Bronkhorst [35] studied the intelligibility and talker recognition against a background of competing voices in monaural and binaural reproduction. They found that the performance between a 3D-auditory display based on HRTFs

and a conventional monaural presentation differs significantly. The binaural presentation yields better speech intelligibility with two or more competing talkers. Yost and colleagues [185] asked participants to identified spoken words, letters, and numbers which were presented in three listening conditions. The three listening conditions are a monaural reproduction of the acoustical scene, a binaural reproduction using headphones and KEMAR-HRTFs for the binaural synthesis and a reproduction over loudspeakers arranged in an anechoic chamber. Results show best performance for "normal listening condition" (reproduction with loudspeakers) and worst for the monaural reproduction. They reasoned that spatial hearing plays an important role in divided attention tasks. Especially, in cocktail-party problems where more than two stimuli are presented simultaneously this effect is found.
The overall conclusion shows that speech intelligibility is better for a spatial scene than for a monaural reproduction. Different than in these investigations, the present paradigm on auditory selective attention focuses especially on the intentional switching of attention [74] and dichotic listening is compared to binaural listening.

4.1.1. Methods

A number of 30 unpaid, student participants aged between 20 and 29 years (mean age: 23.4 ± 2.2 years) participated voluntarily in the experiment. They are equally divided in female and male participants. All of them report that they are normal-hearing and inexperienced in binaural listening tests.

The dichotic-listening paradigm as described in chapter 3.1.1 is used for this experiment and therefore speaker's positions are limited to right and left. For the dichotic reproduction each stimulus is presented to one ear. Headphones as described in chapter 3.7.3 are used for dichotic and binaural reproduction. For the binaural reproduction via headphones individual HRTFs are measured with loudspeakers of type *KH O110D Klein und Hummel* (compare chapter 3.6.2) with microphones placed in earplugs (compare chapter 3.6.1). HpTFs are measured and calculated as reported in chapter 3.6.3. Participants are asked to wear headphones only for dichotic and binaural reproduction. Two equalized loudspeakers of type *KH O110D Klein und Hummel* are used for the reproduction method of real sources (different from those described in chapter 3.7.2). The experiment took place in the anechoic chamber (compare chapter 3.5.1) and stimuli as described in chapter 3.4.1 are used.

There are three independent variables in the present experiment (compare table 4.1). The reproduction method has three levels as described above: dichotic reproduction, binaural reproduction with headphones using individual synthesis, real sources represented by loudspeakers in an anechoic chamber. Attention switch (repetition vs. switch) and congruency (congruent vs. incongruent) are described in chapter 3.2. Dependent variables are reaction times and error rates.

Participants are tested in a within-subject design, including three blocks in total, one of every reproduction method (arranged in Latin square design). Each block includes a training phase of 32 trials and 144 trials used for the analysis. Participants are asked to take a break of 5 minutes between blocks.

Table 4.1.: Experiment I: Independent variables and their levels.

Independent Variables	
Reproduction Method [RM]	Dichotic Binaural Real Sources
Attention Switch [AS]	Repetition Switch
Congruency [C]	Congruent Incongruent

4.1.2. Results

Reaction Times

In reaction times, the ANOVA yields no significant main effect of reproduction method, $[RM\colon F(2,58) = 1.42, p > .05, \eta_p^2 = .05]$.
The main effect of attention switch is significant, $[AS\colon F(1,29) = 43.64, p < .001, \eta_p^2 = .60]$, indicating longer reaction times in switches than in repetitions. The switch costs (compare chapter 2.3.3) amounted on average to 104 ms.
The main effect of congruency is also significant, $[C\colon F(1,29) = 6.96, p = .01, \eta_p^2 = .19]$, indicating longer reaction times in incongruent trials than in congruent trials. No significant interaction can be found (compare table A.1 and A.3). Data is shown in figure 4.1.

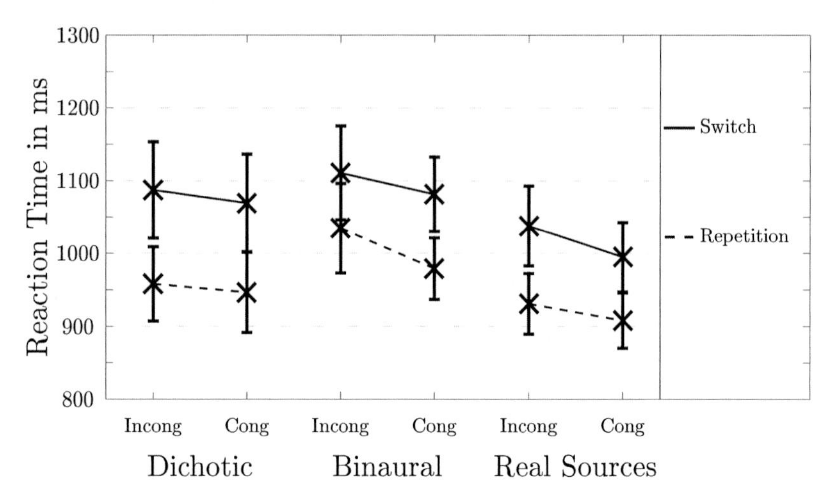

Figure 4.1.: Experiment I: Reaction times (in ms) as a function of reproduction method, attention switch and congruency (RM × AS × C). Error bars indicate standard errors.

Error Rates

In error rates, the same main effects as in RT turn out to be significant. The ANOVA yields no significant main effect of reproduction method, [RM: $F(1.16, 33.69) = 1.44, p > .05, \eta_p^2 = .05$].
The main effect of attention switch is significant, [AS: $F(1, 29) = 30.99, p < .001, \eta_p^2 = .52$], indicating higher error rates in switches than in repetitions.
The main effect of congruency is also significant, [C: $F(1, 29) = 6.26, p < .05, \eta_p^2 = .18$], indicating higher error rates in incongruent trials than in congruent trials. A difference of 3.4 % can be found between congruent and incongruent stimuli. No significant interaction can be found (compare table A.2 and A.3). Data is shown in figure 4.2.

4.1.3. Discussion

Regarding the reproduction method, neither a significant main effect, nor any interaction can be found. Congruency effects and switch costs ([71], compare chapter 3.2.2) agree with findings of earlier experiments [85, 74]. On account of the findings, it can be assumed that this experiment on intentional switching of attention is not highly affected by the reproduction method, but more or less robust.

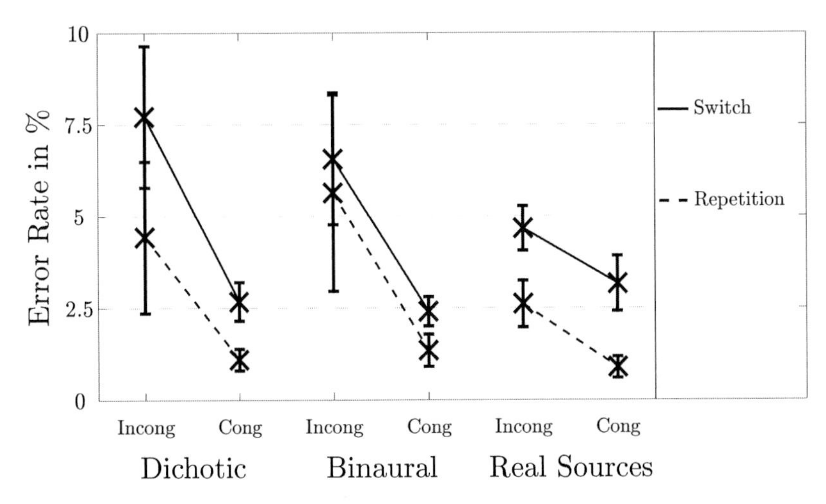

Figure 4.2.: Experiment II: Error rates (in %) as a function of reproduction method, attention switch and congruency (RM × AS × C). Error bars indicate standard errors.

This is in contrast to the results of the named investigations [35, 185], where significantly better results are found when a spatial reproduction is used. Compared to Drullman and Bronkhorst as well as Yost and colleagues a very simple setup, having speakers positioned to the left and right, is used in the present examination. This represents the spatial scene that can be reproduced by a dichotic presentation rather than anything else. The monaural reproduction of a spatial scene, as performed by Yost and colleagues, lacks in imitating the binaural reproduction of the spatial scene. Therefore it is estimated that the quality difference between the monaural and the binaural reproduction is greater than the quality difference between the dichotic and the binaural reproduction in the present simplified setup. Further possible reasons for this disagreement are general differences in the design and the exploratory focus of the experiment.

As a next step the dichotic-listening paradigm is extended to a binaural-listening paradigm in order to investigate auditory selective attention in spatial environments.

4.2. Comparing Binaural Reproduction Methods – Experiment II

Parts of this study are published in Acta Acoustica in 2014 [129]. Experimental data can be downloaded from the technical report [124].

"In simple experimental setups as presented by Koch and colleagues [74] where two sources are presented to the right and left ear, the dichotic reproduction of stimuli is convenient [140, 62]. In general, however, dichotic listening is a highly artificial situation compared to natural listening. A binaural reproduction of stimuli is more natural and offers several advantages"[129].
The binaural-listening paradigm (compare chapter 3.1.2) as a first extension of the dichtoic-listening paradigm (compare chapter 3.1.1) is verified in the present experiment. It is to be ascertained whether the effects of attention switch and congruency yield to the same effects in the binaural setup compared to the dichotic-listening paradigm. Several experiments by Koch, Lawo and colleagues [74, 72, 73, 85, 89, 86] using the dichotic-listening paradigm have shown how the participant's performance declines when the target's location is switched compared to a repetition of the target's position (attention switch, compare chapter 3.2.1). Furthermore, it is expected that the participant's performance of filtering out the irrelevant information delivered by the distracting speaker is deteriorated in incongruent trials compared to congruent trials (congruency, compare chapter 3.2.2) in the binaural-listening paradigm.
The binaural scene presented to the listener in the binaural-listening paradigm can be reproduced by different binaural reproduction methods (compare chapter 3.7). The most straight forward method would be a presentation by loudspeakers positioned around the listener to serve as real sources. Furthermore, binaural stimuli can be created using HRTFs (individual or non-individual) and reproduced via headphones. Cross-Talk Cancellation offers an alternative to headphones for reproducing a binaural synthesis.
"These binaural reproduction methods can differ in accuracy of the binaural synthesis. Evaluations of binaural reproduction methods are usually performed with localization experiments. Comparisons of localization performance between real sources and individual binaural synthesis presented with headphones were analyzed and rated as similar by Bronkhorst [21]. Wightman and Kistler [181] found similar results, but they also report about challenges in elevated positions for the individual binaural synthesis which became apparent through an increased angle of error. The results of comparisons between individual and non-individual binaural recordings were analyzed by several authors [160, 25, 180, 115]. All of them showed that individual recordings yielded better results than non-individual

recordings for localizing sources in space. Detailed results also showed that in localization tasks non-individual binaural stimuli especially caused difficulties for sources located in the median plane, on cones of confusion, as well as in elevated directions"[129].

Using the present binaural-listening paradigm differences, similarities, shortcomings and advantages of the reproduction methods should be validated. It is explored whether the reproduction method has an impact on the listener's performance in general (reaction times and error rates) and whether the performance of intentional attention switching (attention switch and congruency effect) are influenced by the reproduction method.

"With a binaural reproduction, there are more degrees of freedom for the location of sources. [...] In contrast to the limitation to left and right in dichotic listening, in binaural listening, sources can be positioned at any location on a sphere around the listener. [...] Therefore the distance between sources as well as the distance of sources to the listener are more variable [9, 10, 68, 3, 117]. Questions of whether a spatial separation of target and distractor improves attention performance [3, 7] or, whether attention acts like a "spotlight" [8], can only be analyzed and answered with a binaural experimental setup. For example, Bregman [18] and Deutsch [31] reported a benefit of binaural listening emphasizing the ability to switch voluntarily between multiple channels or streams of information"[129].

Based on different localization uncertainties for sources in front, back and to the side [11], it is expected that the listener's performance is affected by the target's position. Localization is to be known as challenging with in the median plane or other cones of confusion [11, 116]. It is expected that the spatial combination of target's and distractor's location makes a contribution to the overall performance of the listener. An interdependence between speaker's arrangement and reproduction methods as well as the performance of intentional attention switching (attention switch and congruency effect) are probable.

4.2.1. Methods

A number of 96 ($4 \cdot 24$, between-subject design) paid students aged between 19 and 34 (mean age: 24.2 ± 3.6 years) participated in the experiment and are randomly assigned to the four reproduction methods. Participants are equally divided into male and female listeners. Listeners are screened to ensure that they have normal hearing (within 20 dB) for frequencies between 250 Hz and 10 kHz. All listeners can be considered as non-expert listeners since they have never participated in a listening test on auditory selective attention.

Different reproduction techniques are compared in the present experiment. The "normal" listening situation is covered by reproduction method *A*: Real sources (compare chapter 3.7.2). Loudspeakers are evenly distributed in an anechoic chamber around the listener. The other reproduction methods are based on binaurally synthesized stimuli. In reproduction method *B*: Individual binaural stimuli via headphones (compare chapter 3.7.3), HRTFs are measured individually. Reproduction method *C*: Non-individual binaural stimuli via headphones and reproduction method *D*: Non-individual binaural stimuli via CTC (compare chapter 3.7.3), both use HRTFs of a dummy head. They differ in the reproduction device. In reproduction method *C* headphones are used, while in reproduction method *D* loudspeakers and a CTC-filter are applied.

The binaural-listening paradigm as described in chapter 3.1.2 is used for this experiment and therefore there are eight possible speaker's positions. Headphones as described in chapter 3.7.3 are used for reproduction methods *B* and *C*. For reproduction methods *A* and *D* loudspeakers as described in chapter 3.7.2 are used. For reproduction method *B* individual HRTFs are measured with loudspeakers arranged in a spherical arrangement (compare chapter 3.6.2) with microphones placed in open-domes (compare chapter 3.6.1). HpTFs are measured and calculated as reported in chapter 3.6.3. Participants are asked to wear headphones only for reproduction methods *B* and *C*. The experiment took place in the anechoic chamber (compare chapter 3.5.1) and stimuli as described in chapter 3.4.1 are used.

There are five independent variables in the present experiment (compare table 4.2). The reproduction method has four levels as described above: *A* - real sources, *B* - individual binaural synthesis reproduced by headphones, *C* - non-individual binaural synthesis reproduced by headphones, *D* - non-individual binaural synthesis reproduced via CTC.
Eight possible positions on the horizontal plane for target- and distracting speaker result in 58 possible spatial combinations. To break this down, the analysis of the spatial combination of target and distractor is split in two variables: Target's position (median vs. diagonal vs. frontal plane), Angle between target and distractor (45 ° vs. 90 ° vs. 135 ° vs. 180 °) (compare chapter 3.2.3 and 3.2.4). Attention switch (repetition vs. switch) and congruency (congruent vs. incongruent) are described in chapter 3.2.1 and 3.2.2. Dependent variables are reaction times and error rates.

Participants are tested in a between-subject design. In total 600 trials divided into four blocks of 150 trials each are separated by short breaks (5 min). The experimental blocks are preceded by one training block of 50 trials. The total duration of the experiment does not exceed 60 min including the audiometry. Trials are counterbalanced over combinations of digits, target's and distractor's postion.

Table 4.2.: Experiment II: Independent variables and their levels.

Independent Variables	
Reproduction Method [RM] (between-subject)	*A* Real Sources *B* Ind. HRTFs via Headphones *C* Non-ind. HRTFs via Headphones *D* Non-ind. HRTFs via CTC
Angle bewteen Target and Distractor [ANG]	45 ° 90 ° 135 ° 180 °
Position of Target [TPOS]	Median Diagonal Frontal
Attention Switch [AS]	Repetition Switch
Congruency [C]	Congruent Incongruent

4.2.2. Results

Main Effects – Reaction Times

For reaction times, all main effects turn out to be significant (compare table A.4). The repeated measures ANOVA yields a significant main effect of reproduction method [RM: $F(3, 92) = 6.67$, $p < .001$, $\eta_p^2 = .18$]. A significant Bonferroni-adjusted post-hoc analysis reveals a significant difference ($p < .01$) in performance of the reproduction method A and D as well as B and D, indicating shorter reaction times ($\sim 1000\,\text{ms}$) for individual reproduction methods (A and B) than for the non-individual reproduction methods ($1100\,\text{ms} - 1200\,\text{ms}$) (compare figure 4.3 and table A.6).
The Huynh-Feldt corrected ANOVA reveals a significant main effect of target's position [$TPOS$: $F(1.66, 152.29) = 49.69$, $p < .001$, $\eta_p^2 = .35$]. A significant post-

hoc analysis determines significant differences ($p < .01$) in performance between all planes, indicating longest reaction times for trials with target positioned on median plane ($\sim$ 1150 ms) and shortest for those on frontal plane ($\sim$ 1050 ms) (compare figure 4.4 and table A.6).
The main effect of the spatial angle between target and distractor is significant [ANG: $F(2.89, 265.92) = 61.69$, $p < .001$, $\eta_p^2 = .40$]. A significant post-hoc analysis determines significant differences ($p < .001$) in performance between the 45° arrangement of target and distractor ($\sim$ 1150 ms) and all other tested angles (1075 ms-1090 ms), indicating longer reaction times for the directly neighbored arrangement of target and distractor (45°) (compare figure 4.4 and table A.6).
The main effect of attention switch on reaction time is significant [AS: $F(1, 92) = 65.18$, $p < .001$, $\eta_p^2 = .42$] and indicates a longer reaction time for switches than for repetitions (compare figure 4.3 and table A.6). The switch costs amounted on average to 53 ms (compare figure 4.3 and table A.6).
The ANOVA also yields a significant main effect of congruency [C: $F(1, 92) = 34.43$, $p < .001$ $\eta_p^2 = .27$], indicating longer reaction times for incongruent stimuli (1125 ms) than for congruent stimuli (1076 ms) (compare figure 4.3 and table A.6).

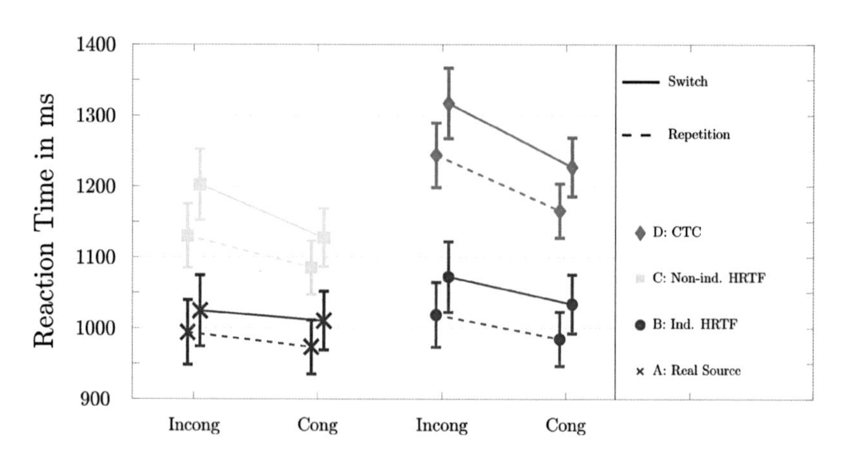

Figure 4.3.: Experiment II: Reaction times (in ms) as a function of reproduction method, attention switch and congruency (RM × AS × C). Error bars indicate standard errors.

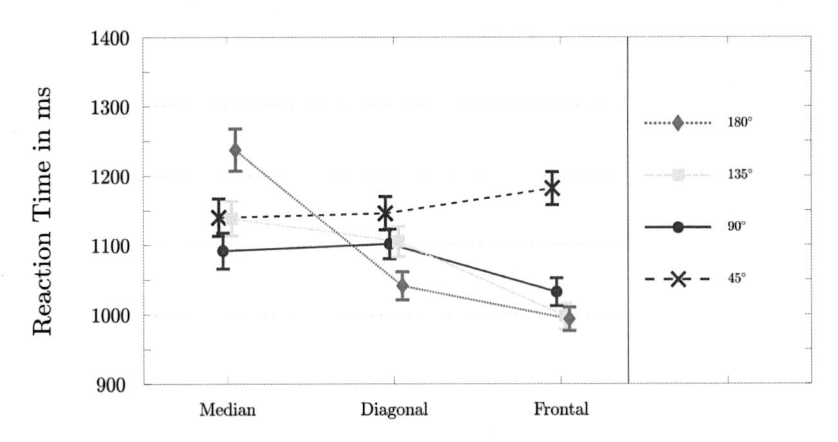

Figure 4.4.: Experiment II: Reaction times (in ms) as a function of position and angle (TPOS × ANG). Error bars indicate standard errors.

Significant Interactions – Reaction Times

The reproduction method interacts with the target's position [$RM \times TPOS$: $F(4.97, 152.28) = 4.01$, $p < .01$, $\eta_p^2 = .12$]. While for reproduction method A the difference between the median plane and the frontal plane amounts to 3 % this difference is 10 % for reproduction method D, respectively (compare figure A.1). The reproduction method also interacts with congruency [$RM \times C$: $F(3, 92) = 2.96$, $p = .04$, $\eta_p^2 = .09$]. The congruency effect is smallest for reproduction method A (congruency effect: 17 ms), is continuously growing with numbering of reproduction methods and is largest for reproduction method D (congruency effect: 84 ms) (compare figure 4.3).

The target's position interacts with the spatial angle between target and distractor [$TPOS \times ANG$: $F(3.89, 357.86) = 54.93$, $p < .001$, $\eta_p^2 = .37$]. The interaction can be seen in figure 4.4. If target and distractor are in a 45 ° arrangement, reaction times are longest for the target being positioned in frontal plane. For all other arrangements reaction times are shortest if the target speaker is positioned in frontal plane. A separation of 180 ° entails short reaction times in frontal and diagonal plane. However, in median plane (meaning target and distractor are positioned in front and back) reaction times are clearly longest for all combinations (1237 ms vs. mean($RT_{180°}$) = 1088 ms).

The three-way interaction with reproduction method turns also out to be significant [$TPOS \times ANG \times RM$: $F(11.67, 357.86) = 2.03$, $p = .02$, $\eta_p^2 = .06$], showing how the effects of the 180 ° arrangement in median plane and the 45 ° arrange-

ment in frontal plane are differently distinct in reproduction methods (compare figure A.1).
The spatial angle between target and distractor interacts with the attention switch [$ANG{\times}AS$: $F(2.87, 263.86) = 5.21$, $p = .002$, $\eta_p^2 = .05$]. If target and distractor are in a 45° arrangement the switch costs add up to 82 ms. For all other angles the switch costs amount to 37 ms - 44 ms (compare figure A.2).
The three-way interaction with reproduction method turns also out to be significant [$ANG{\times}AS{\times}RM$: $F(8.60, 263.86) = 2.24$, $p = .03$, $\eta_p^2 = .07$]. Greatest differences can be observed between reproduction method A and D. While in reproduction method A switch costs decrease with growing angles and vanish for an arrangement angle of 180°, in reproduction method D switch costs stay nearly constant for all angular arrangements, respectively (compare figure A.2).
The target's position interacts with the congruency effect [$TPOS{\times}C$: $F(1.88, 172.67) = 11.01$, $p < .001$, $\eta_p^2 = .11$]. The congruency effect in median plane amounts to 77 ms. For frontal plane and diagonal plane, it is less distinct and amounts to 46 ms and 25 ms, respectively (compare figure A.3).
The three-way interaction with reproduction method turns also out to be significant [$TPOS{\times}C{\times}RM$: $F(5.63, 172.67) = 2.64$, $p = .02$, $\eta_p^2 = .08$], showing how the distinct congruency effect in median plane holds only true for the non-individual reproduction methods (C and D) (compare figure A.3).
Please find further interactions in table A.4.

Main Effects – Error Rates

In error rates, all main effects, except attention switch turn out to be significant (compare table A.5).
The repeated measures ANOVA yields a significant main effect of reproduction method [RM: $F(3, 92) = 17.93$, $p < .001$, $\eta_p^2 = .37$]. A significant Bonferroni-adjusted post-hoc analysis reveals a significant difference ($p < .01$) in performance of the reproduction method A and all other reproduction methods, indicating shorter reaction times for reproduction method A (3.3 %) than for the other reproduction methods (7.5 % - 9.0 %).
The Huynh-Feldt corrected ANOVA reveals a significant main effect of target's position [$TPOS$: $F(1.82, 167.07) = 39.42$, $p < .001$, $\eta_p^2 = .30$] (compare figure 4.6). A significant post-hoc analysis determines significant differences ($p < .01$) in performance between all planes, indicating longest reaction times for trials with target positioned on median plane (9.0 %) and lowest for those on frontal plane (5.6 %).
The main effect of the spatial angle between target and distractor is significant

[ANG: $F(2.84, 261.01) = 81.69$, $p < .001$, $\eta_p^2 = .47$] (compare figure 4.6). A significant post-hoc analysis determines significant differences ($p < .001$) in performance between the 45 ° arrangement and the arrangements of 90 ° and 135 °. As well as, a significant differences ($p < .001$) in performance between the 180 ° arrangement and the arrangements of 90 ° and 135 °.
The main effect of attention switch on reaction time turns out not to be significant [AS: $F(1, 92) = 2.38$, $p = .13$, $\eta_p^2 = .03$].
The ANOVA also yields a significant main effect of congruency [C: $F(1, 92) = 430.21$, $p < .001$, $\eta_p^2 = .82$], indicating longer reaction times for incongruent stimuli than for congruent stimuli (compare figure 4.3 and table A.6). The congruency effect amounts to 9.5 %.

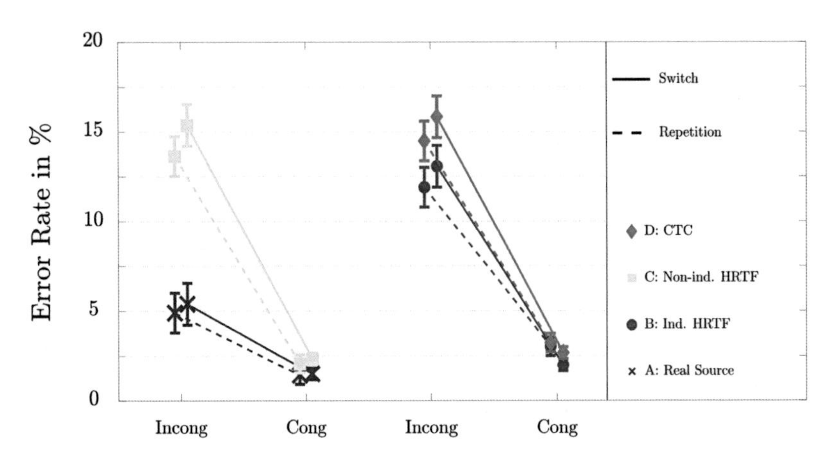

Figure 4.5.: Experiment II: Error rates (in %) as a function of reproduction method, attention switch and congruency (RM × AS × C). Error bars indicate standard errors.

Significant Interactions – Error Rates

The reproduction method interacts with the target's position [$RM \times TPOS$: $F(5.45, 167.07) = 17.93$, $p = .001$, $\eta_p^2 = .37$]. While for reproduction method A the difference between the median plane and the frontal plane amounts to 8.6 %, this difference is 45.0 % for reproduction method D, respectively (compare figure A.4).
The reproduction method interacts with the spatial angle between target and distractor [$RM \times ANG$: $F(8.51, 261.01) = 3.21$, $p = .001$, $\eta_p^2 = .10$]. Generally, error rates are lowest in the 135 ° arrangement for all reproduction methods and highest for the 45 ° arrangement (except reproduction method D: 45 °: 11.7 %;

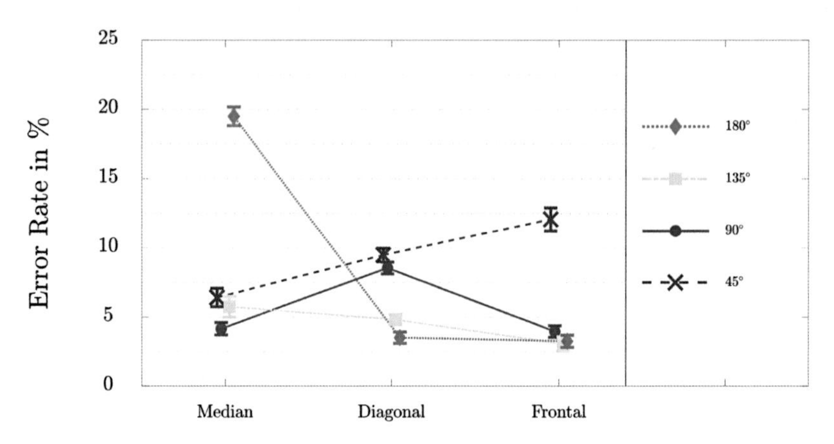

Figure 4.6.: Experiment II: Error rates (in %) as a function of position and angle (TPOS × ANG). Error bars indicate standard errors.

180 °: 11.9 %). Percentage differences between error rates of angular arrangements differ between reproduction methods (compare figure A.4).
The reproduction method also interacts with congruency [$RM \times C$: $F(3, 92) = 19.37$, $p < .001$, $\eta_p^2 = .39$]. The congruency effect is smallest for reproduction method A (congruency effect: 3.6 %), is continuously growing with reproduction methods and is largest for reproduction method D (congruency effect: 12.3 %) (compare figure 4.5).
The target's position interacts with the spatial angle between target and distractor [$TPOS \times ANG$: $F(4.42, 406.49) = 134.90$, $p < .001$, $\eta_p^2 = .60$]. The interaction can be seen in figure 4.6. If target and distractor are in a 45 ° arrangement, error rates are highest for the target being positioned in frontal plane. For all other arrangements error rates are shortest if the target speaker is positioned in frontal plane. A separation of 180 ° entails low error rates in frontal and diagonal plane. However, in median plane error rates are clearly highest for all combinations (19.5 % vs. mean($ErrorRate_{180^\circ}$) = 5.9 %). Target and distractor are positioned on a cone of confusion, where it is more difficult to distinguish between sources, when spatially arranged in 90 ° on diagonal plane. Error rates reflect this with higher values (8.5 %) than for median and frontal plane (mean($ErrorRate_{90^\circ}(median, frontal)$) = 4.1 %).
The three-way interaction with reproduction method turns also out to be significant [$TPOS \times ANG \times RM$: $F(13.26, 406.49) = 4.40$, $p < .001$, $\eta_p^2 = .39$], showing how the effects of the 180 ° arrangement in median plane, the 45 ° arrangement in frontal plane and the 90 ° arrangement in diagonal plane are differently distinct

in reproduction methods (compare figure A.4).
The target's position interacts with the congruency effect [$TPOS{\times}C$: $F(1.68, 154.67) = 36.66$, $p < .001$, $\eta_p^2 = .29$]. The congruency effect in median plane amounts to 13.3 %. For frontal plane and diagonal plane, it is less distinct and amounts to 7.2 % and 8.2 %, respectively (compare figure A.5).
The three-way interaction with reproduction method turns also out to be significant [$TPOS{\times}C{\times}RM$: $F(5.04, 154.67) = 3.89$, $p = .002$, $\eta_p^2 = .11$], showing how the distinct congruency effect in median plane holds only true for the reproduction methods B, C and D (compare figure A.5).
The spatial angle between target and distractor interacts with the congruency effect [$ANG{\times}C$: $F(2.77, 254.53) = 66.25$, $p < .001$, $\eta_p^2 = .42$]. The congruency effect differs significantly for tested arrangements of target and distractor (45 °: 14.2 %; 90 °: 7.2 %; 135 °: 4.5 %; 180 °: 12.3 %) (compare figure A.6).
The three-way interaction with reproduction method turns also out to be significant [$ANG{\times}C{\times}RM$: $F(8.30, 254.53) = 2.95$, $p = .003$, $\eta_p^2 = .09$], showing how the congruency effect dependent on angular arrangement differs between reproduction methods, especially reproduction method A compared to the other reproduction methods (compare figure A.6).
Please find further interactions in table A.5.

4.2.3. Discussion

The dichotic-listening paradigm [74] is extended to a more realistic binaural-listening paradigm in the present experiment. A significant effect of attention switch in reaction times and a significant effect of congruency in reaction times and error rates is observed. Present results are compared to results collected with the dichotic-listening paradigm by Koch, Lawo and colleagues [74, 72, 73, 85, 89, 86]. Compared to Experiment I delineated by Koch and colleagues [74], reaction times are generally longer in the present experiment (dichotic: 1019 ms vs. binaural: 1100 ms). However, error rates are smaller for a binaural reproduction (dichotic: 8.6 % vs. binaural: 7.1 %). The present experiment yields the same significant main effects and interactions in attention switch and congruency (compare table A.4 and table A.5).
The significant effect of attention switch in reaction times, indicating that subjects responded more slowly when the target's direction is switched, can be observed in all performed experiments [40, 127, 129, 74, 72, 73, 85, 89, 86]. The switch cost provide an explicit measure of how well instructions to switch attention could be followed. In reaction times, switch costs are lower in the binaural experiment compared to the dichotic experiment (dichotic: 126 ms, 12.4 % compared to the mean reaction times vs. binaural: 52 ms, 4.7 % compared to the mean reaction

times). "A switch between only two possible directions (i.e. dichotic listening) [...] [is] expected to be easier to detect than a switch to one of eight possible directions equally distributed on the horizontal plane. The angular distance between the target's positions could [...] [be] a reason for different complexity of switches. Besides the angular distance of target's positions, the visual cue (in ear-based dichotic experiments the visual cue [...] [is] a letter (L/R) and therefore [...] [differs] from the cue design of this investigation) as well as the reproduction method (binaural vs. dichotic) might have [...] an effect on the switch costs"[129]. "While the significant effect of congruency [...] [is] less apparent for reaction times, the difference between congruent and incongruent trials [...] [is] more pronounced for error rates. These findings [...] [are] also confirmed by the previous dichotic investigations [74, 72, 73, 85, 89, 86]. The congruence effect [...] [can] be taken as an implicit performance measure of attending to task-irrelevant information and filtering out the irrelevant information [74]"[129]. The congruency effect in error rates is greater in the binaural experiment even though error rates are generally lower (dichotic: 5.4 %, 62.8 % compared to the mean error rates vs. binaural: 9.5 %, 133.8 % compared to the mean error rates).

One important research question of this experiment is whether the four binaural reproduction methods differ by means of a paradigm focusing on intentional switching in auditory selective attention. "There [...] [is] statistical evidence that absolute values of reaction times and error rates differed between reproduction methods. By contrast with other investigations that compare reproduction methods in localization experiments, similarities and differences [...] [can] be found"[129].
For reproduction method A (Real Sources) lowest error rates and shortest reaction times are found. Reproduction method A does not differ significantly from reproduction method B in reaction times, as expected since individual HRTFs are used to present binaural stimuli. However, in error rates the two reproduction methods differ significantly. Supposedly, the difference between reproduction method A and B is due to the static presentation of the binaural synthesis in reproduction method B. "In both reproduction methods [...] [participants are] able to perform small head movements (sounding) [[66, 143, 177]] within the permitted area defined by the tracker. While [...] [participants] listening to real sources [...] [get] a feedback in terms of changes in interaural level difference (ILD) and interaural time difference (ITD) from the movements of sounding, [...] [participants] listening to the binaural synthesis miss this additional localization information. The static presentation of individual binaural stimuli [...] [does] not offer the additional localization information of head movements and therefore, it [...] [can] be assumed that error rates were increased at least partly due to the

lack of this advantage"[129].
In a localization experiment Bronkhorst [21] found that participants localize almost equally accurate when receiving sounds reproduced by headphones based on an individualized binaural synthesis compared to real sources. In contrast to the present experiment, in Bronkhorst's experiment it is provided that head movements can be made. Moller and colleagues [116] agree even though no head movements are allowed in their experiment. They state: "When compared to real life, the localization performance was preserved with individual recordings"[116]. It became general knowledge that non-individual binaural synthesis lacks in localization quality compared to individual reproduction methods [116, 21, 180]. In the present experiment no significant difference between reproduction method B and C, as well as method A and C in reaction times and error rates can be found. A tendency towards longer reaction times and higher error rates for non-individual binaural reproduction using headphones can be observed compared to the individual binaural reproduction methods (A and B). Reproduction method D, which is also based on non-individual data achieve the results with longest reaction times and highest error rates. A significant difference between reproduction method A and D in reaction times and error rates and a significant difference between B and D in reaction times is found. Since, no significant difference between the individual and the non-individual reproduction method (B and C) is found, the present difference is also based on the reproduction device (headphones vs. CTC). "In CTC evaluations [47, 167, 92, 5] concerning localization, limited sweet spots raise a challenge and affect performance. Since head movements are supervised limitation due to the sweet spot do not severely affect the results of this reproduction method in the present study"[129]. However, a CTC-reproduction using only two loudspeakers is only stable within the spanned angle of the loudspeakers. An extension to four loudspeakers is a more robust solution for a dynamic CTC-reproduction [47, 90].

Significant differences in absolute values of reaction times and error rates for the tested reproduction methods are found. Furthermore, it is examined whether the different reproduction methods have any impact on the effects of auditory attention switching and congruency. While the effect of attention switching is not influenced by the reproduction methods, the reproduction methods largely affected the congruency effect. Participants are more challenged to suppress the interfering speech when the binaural reproduction is based on non-individual HRTFs. This effect is found in reaction times and error rates. In error rates the congruency effect is also differently distinct within individual reproduction methods, indicating the smallest congruency effect when using real sources.

The new binaural-listening paradigm offers eight possible source locations on the horizontal plane and entails further effects regarding the spatial position and angular combination of target and distractor. It is observed how the participants performance (RT and ER) drops significantly when target and distractor are located on the median plane (meaning target and distractor are positioned in front and back) compared to the frontal plane. This effect is supposedly based on different degrees of difficulty in localization especially in the extreme cases of median plane (ITD and ILD information vanish) and frontal plane (ITD and ILD information are maximal). When using CTC-reproduction the effect in median plane is even more pronounced than in the binaural reproduction methods A and B. Localization experiments confirm these findings. For example Møller and colleagues [116] found accumulated errors in "Median Plane" and "Within Cone" conditions which is especially true for non-individual reproduction methods. In agreement, Wenzel [180] also found increased median plane errors when using non-individual HRTFs.
Generally, highest error rates and longest reaction times are observed when target and distractor are in a 45 ° arrangement and therefore directly neighbored. The effect of spatial separation of sources is also studied by Best and colleagues [8] in an experiment focusing on selective attention. Observing error rates, it is shown that auditory selective attention is exposed to a greater challenge when sources are not or only little spatially separated.
Accuracy of localization in the horizontal plane is to be known as best in front and worst to the sides [11]. It is reasoned that the effect that reaction times and error rates rise when the target is being positioned to the side and the distractor is positioned directly next to the target on diagonal plane is due to the lack of the localization ability. This effect is even more pronounced in the non-individual reproduction methods.

Another benefit of the binaural extension of the paradigm is that the performance of intentional attention switching can be analyzed depending on the speakers' positions. The participants ability to switch attention declines when target and distractor are next neighbors (45 °). It is assumed that switching the attention to a new position which is directly neighbored by a distracting source, gives more uncertainty whether the correct source is focused.
Suppressing the distractor's speech is significantly more difficult when the target is positioned on median plane and even more if the distractor is also located on median plane and the reproduction method is based on a binaural synthesis relative to the real sources. The loss of the additional localization information due to a static reproduction can be a reason for this effect. The congruency effect is also influenced when the speech sources are directly neighbored. When using

non-individual reproduction methods this effect is even more distinct. From a spatially acoustically point of view these challenging arrangements of target and distractor are self-explanatory. The effects in absolute reaction times and error rates can be transferred to the congruency effect.

According to this investigation, the extension of a dichotic-listening paradigm to a binaural-listening paradigm affords greater opportunities to analyze intentional switching in auditory attention. The arrangement of target and distractor have a great impact on the cognitive processing, especially in suppressing the attention towards the distracting sources. It is therefore inevitable to analyze these relations in further experiments. However, the amount of data taken in some subsequent experiments is limited which is why the angular arrangement of target and distractor is disregarded in several cases.
The comparison of reproduction methods showed that the differences between absolute values of reaction time and error rates should not be neglected. The loss of individual information (reproduction methods C and D) and the restriction of head movements (reproduction methods B, C and D) diminish the ability of ignoring a distracting speaker in a spatial setup and therefore intensifies the observed effect of congruency. The performance of attention switching is negligibly affected by the reproduction methods.
Reproduction method A, using real sources, offers results most similar to those collected before in experiments using the dichotic-listening paradigm. However, this reproduction method is inconvenient, since an anechoic chamber is necessary. Furthermore, it is not possible to examine reverberation effects without setting the experiment in reverberating rooms. Using a binaural synthesis augmented by a room model and the belonging acoustic computations, a large number of reverberation times in different rooms can be tested. Since no difference in performance of cognitive processing (congruency effect and attention switch) between the reproduction methods B (individual HRTFs) and C (non-individual HRTFs) presented via headphones is found, it is decided to use the binaural reproduction method based on non-individual HRTFs reproduced via headphones (C) for subsequent experiments. In terms of applicability, it is the most convenient and feasible reproduction method. Experiments can be performed in hearing booths instead of an anechoic chamber and no individual HRTFs need to be measured.
As a next step towards auditory selective attention in realistic scenes, the listener is set into scenarios taken place in rooms of different reverberation.

4.3. Reverberation – Constraints of the Binaural-Listening Paradigm – Experiment III

Parts of this study are presented at the national conference on acoustics DAGA in 2014 [128]. Experimental data can be downloaded from the technical report [124].

As a next step towards realistic cocktail-party scenes reverberation is included in the auralization [128]. In real-life scenes reverberant energy distorts the signal [122, 28, 84] and therefore it is also of interest how auditory selective attention is affected by reverberant energy.
"Using an attention task where [participants][...] are asked to repeat four consecutive digits spoken by the target speaker always positioned in front in the presence of two other distracting speakers located to the sides, Ruggles and Shinn-Cunningham [152] varied the amount of reverberant energy from "anechoic ($RT_{60} = 0\,\mathrm{s}$), intermediate reverberation ($RT_{60} = 0.4\,\mathrm{s}$) to high reverberation ($RT_{60} = 3\,\mathrm{s}$)". They reported a great impact on performance when adding reverberation, especially differences in performance between anechoic (60-80% correct) and intermediate reverberation (40-50% correct) are noteworthy. On account of these results they conclude that reverberant energy interferes with spatial selective attention.
Similar reverberation times are analyzed by Culling and colleagues [27] who measured Speech Reception Thresholds under anechoic ($RT_{60} = 0\,\mathrm{s}$) and reverberant ($RT_{60} = 0.4\,\mathrm{s}$) conditions. Target and distractor are collocated in front of the [participant][...] or spatially separated ($-60\,°/+60\,°$). Speech Reception Thresholds are found to be significantly lower under anechoic conditions, which is reconfirmed by Lavandier and Bronkhorst [83]. The reverberant energy also interacts with the location of target and distractor, indicating no improvement in Speech Reception Threshold for spatially separated speakers in the reverberant condition.
Contradictory to that are findings by Kidd and colleagues [68]. They reported that the effect of reverberation is greater when target and masker are spatially separated rather than collocated at the same position. Instead of using simulated reverberation times, Kidd and colleagues changed the reverberation of the laboratory by mounting foam and plexiglas to the walls. Further findings are that the amount of masking increased as reverberation times increased and that these acoustic differences also significantly affected the performance in the speech identification task.
Related to the cited investigations [152, 27, 68], Darwin and Hukin [29] explored the effect of reverberation ($RT_{60} = 0.4\,\mathrm{s}$) on the ability of listeners to maintain

their attention to one speaker across time. Using a paradigm with minimal intelligibility requirements it is found that the influences of reverberant energy on inter-aural time differences (ITD) are significant. The use of ITD differences is impaired by reverberation and therefore maintaining attention to the target is more complicated. However, natural prosody and vocal-tract size differences between talkers, being two further cues for selective attention, are not affected by reverberation"[136].

"Inspired by the findings of Ruggles and Shinn-Cunningham [152], reverberation times in three levels from anechoic ($RT_{60} = 0\,\mathrm{s}$), low reverberation ($RT_{60} = 0.8\,\mathrm{s}$, comparable to an acoustically untreated classroom instead of $RT_{60} = 0.4\,\mathrm{s}$, comparable to a damped recording room) to high reverberation ($RT_{60} = 1.75\,\mathrm{s}$, comparable to a auditorium instead of $RT_{60} = 3\,\mathrm{s}$, comparable to a medium-sized church) are simulated (compare chapter 3.8). The underlying room model is also designed with comparable diameters, however, walls are not set to be parallel and the listener is not positioned in the center of the room to prevent unwanted acoustical effects due to nodal points or echos [52, 146, 50]" [136].

"It is postulated that reverberant energy increases reaction times and error rates in the present investigation, based on the cited findings [152, 27, 69, 29]. Furthermore, Ruggles and Shinn-Cunningham [152] showed how maintaining auditory selective attention on a single sound source in presence of interfering sources is degraded by reverberant energy. These findings led to the hypothesis of increased reaction times and error rates for repetition trials (i.e. where a listener is asked to focus on the same direction in two consecutive trials) under increased reverberation in the present investigation. Since it is known, how reverberation degrades ITD timing information, which results in a blurred localization information [152], it is predicted in the present investigation that localizing a new sound source and focusing attention on that source would also degrade with increasing reverberation times. This is the case for switch trials where the listener has to switch his/her attention to a new spatial position between trials.
Spatial separation turned out to be beneficial in findings by Kidd and colleagues [69] under increasing reverberation times, however, Culling and colleagues [27] reported an opposite effect. Therefore, in this investigation special attention is focused on the spatial location of target and distractor as well as their angular separation" [136].

4.3.1. Methods

A number of 48 ($3 \cdot 16$, between-subject design) paid, student participants aged between 18 and 27 years (mean age: 22.3 ± 2.3 years) took part in the experiment. They were equally divided in female and male participants. Listeners are screened to ensure that they had normal hearing (within 20 dB) for frequencies between 250 Hz and 10 kHz. All listeners can be considered as non-expert listeners since they have never participated in a listening test on auditory selective attention.

The binaural-listening paradigm as described in chapter 3.1.2 is used for this experiment and therefore stimuli presenting only numbers are applied (compare chapter 3.4.1). Headphones as described in chapter 3.7.3 are used for all participants. Binaural synthesis is based on HRTFs of the dummy head, but headphones are equalized by individually measured HpTFs (compare chapter 3.6.3). The experiment took place in the darkened hearing booth (compare chapter 3.5.2). Using RAVEN stimuli are adjusted to three different reverberation levels (compare chapter 3.8).

Table 4.3.: Experiment III: Independent variables and their levels.

Independent Variables	
Reverberation [R] (between-subject)	Anechoic Low Reverberation High Reverberation
Position of Target [TPOS]	Median Diagonal Frontal
Attention Switch [AS]	Repetition Switch
Congruency [C]	Congruent Incongruent

There are four independent variables in the present experiment (compare table 4.3). The between-subject variable reverberation has three levels as described above (anechoic vs. low reverberation vs. high reverberation).
Target's position (median vs. diagonal vs. frontal plane), attention switch (repetition vs. switch) and congruency (congruent vs. incongruent) are described in chapter 3.2. Dependent variables are reaction times and error rates.

Participants are tested in a between-subject design. In total 600 trials divided into four blocks of 150 trials each are separated by short breaks (5 min). The experimental blocks are preceded by one training block of 50 trials. The total duration of the experiment does not exceed 60 min including the audiometry. Trials are counterbalanced over combinations of repetitions and switches, target's and distractor's postion.

4.3.2. Results

Reaction Times

In reaction times, the repeated measures ANOVA yields no significant main effect of the between-subject variable reverberation time [R: $F < 1$, $\eta_p^2 = .01$].
The Huynh-Feldt corrected ANOVA reveals a significant main effect of target's position [$TPOS$: $F(1.58, 70.95) = 54.46$, $p < .001$, $\eta_p^2 = .55$] (compare figure 4.7). A significant post-hoc analysis determines significant differences ($p < .002$) in performance between all planes, indicating longest reaction times for trials with target positioned on median plane and lowest for those on frontal plane (compare table A.9).
The main effect of attention switch on reaction time is significant [AS: $F(1, 45) = 40.28$, $p < .001$, $\eta_p^2 = .47$] and indicates a longer reaction time for switches than for repetitions (compare figure 4.7). The switch costs amounted on average to 54 ms.
The ANOVA also yields a significant main effect of congruency [C: $F(1, 45) = 13.21$, $p = .001$ $\eta_p^2 = .23$], indicating longer reaction times for incongruent stimuli than for congruent stimuli (compare figure A.7 and table A.9).

There are four significant interactions in reaction times. The interaction of target's position and attention switch turns out to be significant [$TPOS \times AS$: $F(1.79, 80.52) = 4.65$, $p = .02$ $\eta_p^2 = .09$], indicating significantly different switch costs for varying target's positions (switch costs: median plane 81 ms, diagonal plane 38 ms, frontal plane 42 ms).
The interaction of target's position and congruency is significant [$TPOS \times C$: $F(2, 90) = 8.28$, $p < .001$ $\eta_p^2 = .16$], indicating a significantly different congruency effect for varying target's positions (congruency effect: median plane 60 ms, diagonal plane 55 ms, frontal plane 11 ms, also compare figure A.7).
The triple interaction of target's position, attention switch and congruency turns out to be also significant [$TPOS \times AS \times C$: $F(2, 90) = 6.56$, $p = .002$ $\eta_p^2 = .10$]. Finally, the interaction of all variables is just significant [$TPOS \times AS \times C \times R$: $F(4, 90) = 2.51$, $p = .047$ $\eta_p^2 = .10$] (compare figure A.7).

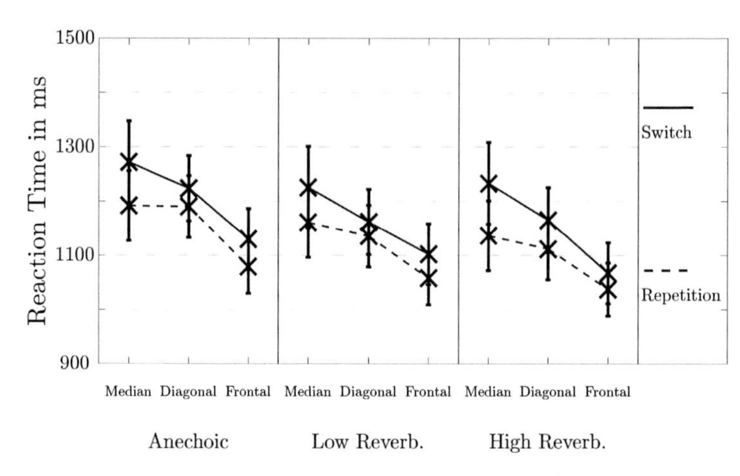

Figure 4.7.: Experiment III: Reaction times (in ms) as a function of reverberation, target's position and attention switch (R × TPOS × AS). Error bars indicate standard errors. Note that for the sake of clarity the variable of congruency is not visualized in this plot. figure A.7 shows all four variables.

Error Rates

In error rates, the repeated measures ANOVA yields no significant main effect of the between-subject variable reverberation time [R: $F(2,45) = 2.68$, $p = .08$, $\eta_p^2 = .11$].
The ANOVA reveals a significant main effect of target's position [$TPOS$: $F(2,90) = 42.14$, $p < .001$, $\eta_p^2 = .48$] (compare figure 4.8). A significant post-hoc analysis determines significant differences ($p < .001$) in performance between frontal plane and the other two planes, indicating highest error rates for trials with target positioned on median plane and lowest for those on frontal plane (compare table A.9).
The main effect of attention switch on reaction time is not significant [AS: $F(1,45) = 2.27$, $p = .14$, $\eta_p^2 = .05$].
The ANOVA yields a significant main effect of congruency [C: $F(1,45) = 271.39$, $p < .001$ $\eta_p^2 = .86$], indicating higher error rates for incongruent stimuli than for congruent stimuli (compare figure 4.8 and table A.9). The congruency effect amounts to 12.4 %.

There are two significant interactions in error rates. The interaction of target's position and congruency turns out to be significant [$TPOS \times C$: $F(2,90) = 38.41$,

$p < .001$ $\eta_p^2 = .45$], indicating that the congruency effect differs for the three planes of target's position. The congruency effect on median and diagonal plane is significantly different from the congruency effect on frontal plane (congruency effect: median plane 16.6 %, diagonal plane 13.4 %, frontal plane 7.5 %, also compare figure 4.8).

Furthermore, the interaction of attention switch and congruency turns out to be significant [$AS \times C$: $F(1, 45) = 6.47$, $p = .02$ $\eta_p^2 = .13$]. Switch costs in congruent trials compared to those in incongruent trials differ significantly (switch costs: congruent −0.3 % vs. incongruent 1.5 %, also compare figure A.8).

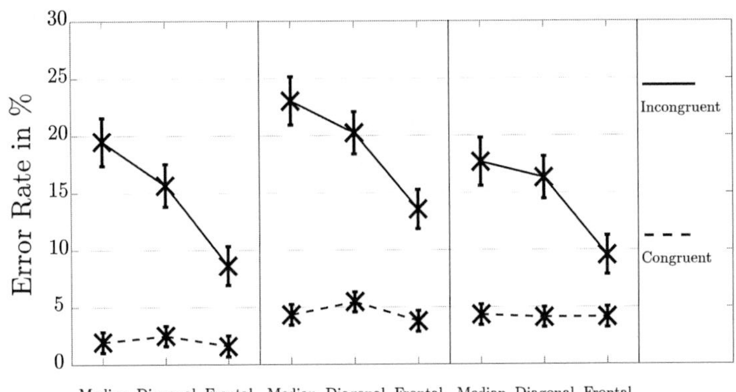

Figure 4.8.: Experiment III: Error rates (in %) as a function of reverberation, target's position and congruency (R × TPOS × C). Error bars indicate standard errors. Note that for the sake of clarity the variable of attention switch is not visualized in this plot. figure A.8 shows all four variables.

4.3.3. Discussion

Regarding the reverberation time, neither a significant main effect, nor any interaction can be found (besides the 4-way interaction). This is in contrast to the postulation and the results of the named investigations [152, 27, 69, 29] who found significantly worse results when participants are tested in reverberant environments.

Compared to Ruggles and Shinn-Cunningham [152], very short stimuli of 730 ms (only one digit compared to a set of four digits, compare chapter 3.4.1) are used in the present examination. Besides the short duration, stimuli are also monosyllabic except for the digit “seven”. Long reverberation times entail the superposition of late reverberation of a syllable and the presentation of the following syllables. Ruggles and Shinn-Cunningham report about a “buildup of spatial selective auditory attention”, where in anechoic and low reverberation conditions the percent correct of digits is increasing within the digit-sequence and in high reverberation conditions it remains constant. Therefore it is believed that no significant difference between reverberation times can be found because of the short monosyllabic stimuli. Furthermore, the used design of investigating reverberation as a between-subject variable could be a restriction of finding significant differences between reverberation times.
As a next step the binaural-listening paradigm is extended (compare chapter 3.1.2 - 3.1.3) in order to revise auditory selective attention in reverberant environments using longer polysyllabic stimuli (compare chapters 4.4 - 4.5).

4.4. Extension to New Binaural-Listening Paradigm – Experiment IV

Parts of this study are presented at the international conference on acoustics ICA in Buenos Aires in 2016 [43] and published in the proceedings POMA in 2017 [44]. Experimental data can be downloaded from the technical report [124].

When analyzing reverberation it turned out to be a shortcoming of the binaural-listening paradigm that stimuli are rather short and monosyllabic. To be able to examine auditory selective attention in realistic, and therefore also reverberant environments, the binaural-listening paradigm is extended into a binaural-listening paradigm with more complex and longer stimuli [43, 44].
The extended stimulus presented by target and distracting speaker are composed of a German digit and a German dissyllabic direction word ("UP", in German "OBEN" and "DOWN", in German "UNTEN"). Stimuli have a duration of 1200 ms (compare chapter 3.4.2).
The extended binaural-listening paradigm is examined and compared to the former binaural-listening paradigm with respect to cognitive performance.

4.4.1. Methods

A number of 48 paid ($2 \cdot 24$, between-subject design), student participants aged between 20 and 32 years (mean age: 24.6 ± 3.0 years) take part in the experiment. They are equally divided in female and male participants. Listeners are screened to ensure that they have normal hearing (within 20 dB) for frequencies between 250 Hz and 10 kHz. All listeners can be considered as non-expert listeners since they have never participated in a listening test on auditory selective attention.

Participants are equally split into participants that perform in the binaural-listening paradigm as described in chapter 3.1.2 and those who test the extended binaural-listening paradigm (compare chapter 3.1.3). Since stimuli are assigned to the used paradigm. Participants tested in the former binaural-listening paradigm listened to stimuli as described in chapter 3.4.1 and those who are tested with the extended binaural-listening paradigm have to evaluate the longer stimuli (compare chapter 3.4.2).
Headphones as described in chapter 3.7.3 are used for all participants. Binaural synthesis is based on HRTFs of the dummy head, but headphones are equalized by individually measured HpTFs (compare chapter 3.6.3). The experiment took place in the darkened hearing booth (compare chapter 3.5.2).

There are four independent variables in the present experiment (compare table 4.4). The between-subject variable of paradigm has two levels as described above (former binaural-listening paradigm vs. extended binaural-listening paradigm). Target's position (median vs. diagonal vs. frontal plane), attention switch (repetition vs. switch) and congruency (congruent vs. incongruent) are described in chapter 3.2. Dependent variables are reaction times and error rates.

"The congruency effect [...] [has] to be redefined. The [...] [participants' task is] still to categorize the relevant digit presented by the target speaker as smaller or larger than five and press the corresponding response button. These categories [...] [are] mapped to the left hand buttons and the right hand buttons at the front side of a controller. Furthermore, the direction word presented by the target [...] [gives] information whether the index finger (in case the direction word [...] [is] "UP") or the middle finger (in case the direction word was "DOWN"). Therefore, four response possibilities were given in a quadratic arrangement to be pressed by index fingers and middle fingers of both hands" [44].

Participants are tested in a between-subject design. In total 600 trials divided into four blocks of 150 trials each are separated by short breaks (5 min). The experimental blocks are preceded by one training block of 50 trials. The total duration of the experiment does not exceed 60 min including the audiometry. Trials are counterbalanced over combinations of digits, target's and distractor's postion.

Table 4.4.: Experiment IV: Independent variables and their levels.

Independent Variables	
Paradigm [P] (between-subject)	Former New
Position of Target [TPOS]	Median Diagonal Frontal
Attention Switch [AS]	Repetition Switch
Congruency [C]	Congruent Incongruent

4.4.2. Results

Reaction Times

In reaction times, the repeated measures ANOVA yields no significant main effect of the between-subject variable paradigm [P: $F(1,46) = 1.04$, $p = .31$, $\eta_p^2 = .02$]. The Huynh-Feldt corrected ANOVA reveals a significant main effect of target's position [$TPOS$: $F(1.40, 64.36) = 58.95$, $p < .001$, $\eta_p^2 = .56$] (compare figure 4.9). A significant post-hoc analysis determines significant differences ($p < .001$) in performance between all planes, indicating longest reaction times for trials with target positioned on median plane and lowest for those on frontal plane (compare table A.12).
The main effect of attention switch on reaction time is significant [AS: $F(1,46) = 41.14$, $p < .001$, $\eta_p^2 = .47$] and indicates a longer reaction time for switches than for repetitions (compare figure 4.9). The switch costs amounted on average to 78 ms.
The ANOVA also yields a significant main effect of congruency [C: $F(1,46) = 7.31$, $p = .01$ $\eta_p^2 = .14$], indicating longer reaction times for incongruent stimuli than for congruent stimuli (compare figure 4.9 and table A.12).
There are two significant interactions in reaction times. The interaction of target's position and attention switch turns out to be significant [$TPOS \times AS$: $F(1.78, 81.77) = 3.88$, $p = .03$ $\eta_p^2 = .08$], indicating significantly different switch costs for varying target's positions (switch costs: median plane 101 ms, diagonal plane 78 ms, frontal plane 55 ms).
There is a significant triple interaction of target's position, congruency and paradigm [$TPOS \times C \times P$: $F(1.83, 84.43) = 4.80$, $p = .01$ $\eta_p^2 = .10$], indicating that the congruency effect relating to the target's position differs between the former binaural-listening paradigm and the extended binaural-listening paradigm. In median and diagonal plane the congruency effect disappears for the new extended paradigm (Congruency effect for the extended binaural-listening paradigm: median plane 6 ms, diagonal plane 5 ms compared to the congruency effect for the former binaural-listening paradigm: median plane 83 ms, diagonal plane 55 ms, also compare figure 4.9).

Error Rates

In error rates, the repeated measures ANOVA yields a significant main effect of the between-subject variable paradigm [P: $F(1,46) = 36.24$, $p < .001$, $\eta_p^2 = .44$], indicating higher error rates for the new extended binaural-listening paradigm (14 % vs. 8 %) (compare table A.12).
The ANOVA reveals a significant main effect of target's position [$TPOS$:

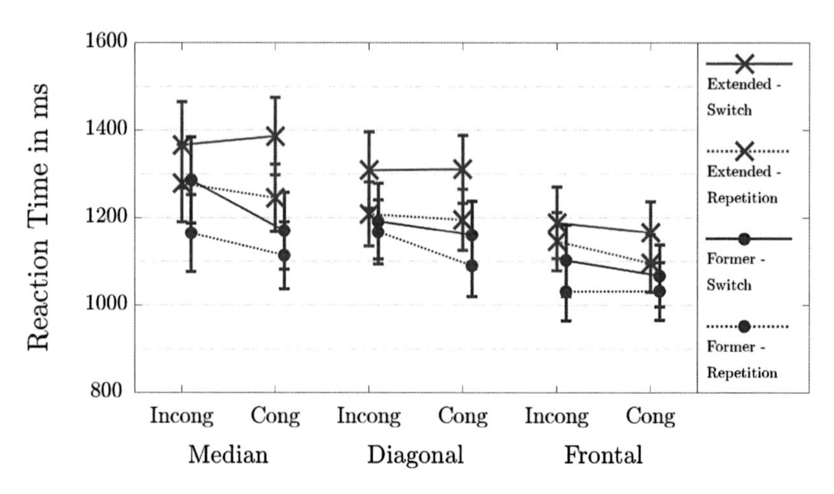

Figure 4.9.: Experiment IV: Reaction times (in ms) as a function of paradigm, target's position, attention switch and congruency (P × TPOS × AS × C). Error bars indicate standard errors.

$F(2,92) = 49.55$, $p < .001$, $\eta_p^2 = .52$] (compare figure 4.10). A significant post-hoc analysis determines significant differences ($p < .001$) in performance between frontal plane and the other two planes, indicating highest error rates for trials with target positioned on median plane and lowest for those on frontal plane (compare table A.12).
The main effect of attention switch on error rates is significant [AS: $F(1,46) = 11.57$, $p = .001$, $\eta_p^2 = .20$] and indicates a higher error rates for switches than for repetitions (compare figure 4.10).
The ANOVA also yields a significant main effect of congruency [C: $F(1,46) = 234.84$, $p < .001$ $\eta_p^2 = .84$], indicating higher error rates for incongruent stimuli than for congruent stimuli (compare figure 4.10 and table A.12). The congruency effect amounts to 11.4 %.

There are two significant interactions in error rates. The interaction of target's position and paradigm turns out to be significant [$TPOS{\times}P$: $F(2,92) = 4.61$, $p = .01$ $\eta_p^2 = .09$]. Error rates in median and diagonal plane compared to those in frontal plane differ significantly for the new extended and the former binaural-listening paradigm. However, the percental difference between frontal plane and the other planes is greater for the new extended paradigm (Extended: $\text{mean}(Error(median, diagonal)) - Error(frontal) = 8.2\,\%$ vs. for-

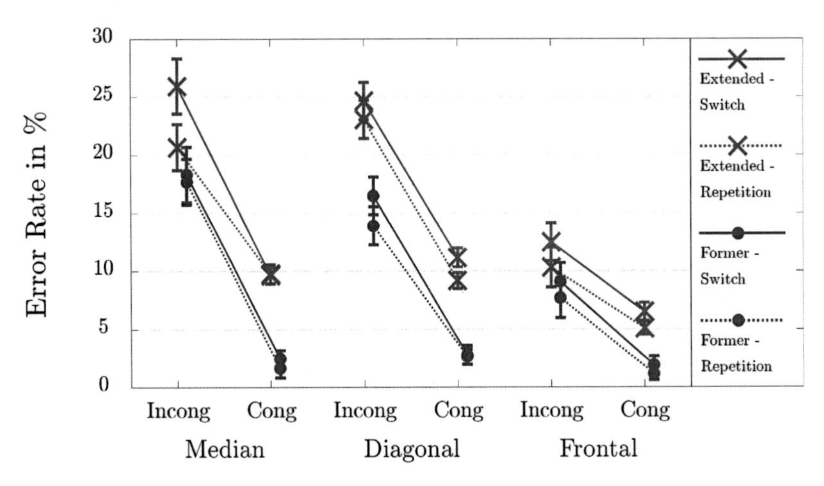

Figure 4.10.: Experiment IV: Error rates (in %) as a function of paradigm, target's position, attention switch and congruency (P × TPOS × AS × C). Error bars indicate standard errors.

mer: mean($Error(median, diagonal)$) – $Error(frontal) = 4.5\,\%$) (compare figure 4.10).
Furthermore, the interaction of Target's Position and Congruency turns out to be significant [$TPOS \times C$: $F(2, 92) = 26.37$, $p < .001$ $\eta_p^2 = .36$], indicating that the congruency effect differs for the three planes of target's position. The congruency effect on median and diagonal plane is significantly different from the congruency effect on frontal plane (Congruency effect: median plane 14.8 %, diagonal plane 13.1 %, frontal plane 6.2 %, also compare figure 4.10).

4.4.3. Discussion

The main aim of this experiment was to verify the newly extended binaural-listening paradigm and compare it to the former binaural-listening paradigm with regard to cognitive performance.
The contrasted paradigms significantly differ in error rates, yielding higher error rates for the extended paradigm. Reaction times are also longer for the extended paradigm, but do not differ significantly. "The increase in reaction time and error rates is reasonable since the answering task is more demanding (four answering possibilities vs. two answering possibilities)"[44].
The significant effect of congruency in reaction times dissolves for the target's position being on the median or diagonal plane when using the new paradigm.

In contrast to that the congruency effect turns out to be significant in reaction times with the former binaural-listening paradigm. It is suggested that the new task of categorizing the digit and rating the direction word is more demanding and masks the effect of congruency in reaction times. In preceding experiments [74, 72, 73, 85, 89, 86, 40, 127, 129] it is always found that the main effect of congruency is more pronounced in error rates, which holds also true for the new extended binaural-listening paradigm.
It is concluded that first results using the extended binaural-listening paradigm show that it is robust and provides comparable findings to the original binaural-listening paradigm. As a next step open research questions on the impact of reverberant energy on cognitive performance (Experiment III) are transferred to the new extended binaural-listening paradigm.

4.5. Steps towards Realistic Environments – Reverberation – Experiment V

Parts of this study are presented at the meeting of the acoustical society of America ASA in 2017 [133] and published in Hearing Research in 2018 [136]. Experimental data can be downloaded from the technical report [124].

In real-life scenes reverberant energy distorts the signal of target and distracting sources [122, 28, 84]. It is therefore of interest how auditory selective attention is affected by reverberant energy. Since when testing with the former binaural-listening paradigm no effect is found (compare chapter 4.3, [128]) a new attempt [133, 136] is made using the extended binaural-listening paradigm utilizing longer, polysyllabic stimuli (compare chapter 4.4, 3.1.3, 3.4.2, [43, 44]).
As discussed in detail in chapter 4.3 an increase of reaction times and error rates dependent on the amount of reverberant energy is expected.

4.5.1. Methods

A number of 24 paid, student participants aged between 19 and 34 years (mean age: 23.9 ± 3.4 years) take part in the experiment. They are equally divided in female and male participants. Listeners are screened to ensure that they have normal hearing (within 20 dB) for frequencies between 250 Hz and 10 kHz. All listeners can be considered as non-expert listeners since they have never participated in a listening test on auditory selective attention. One participant had to be excluded from the analysis due to missing data in reaction times.

The extended binaural-listening paradigm as described in chapter 3.1.3 is used for this experiment, as a reissue of Experiment III. Different to the first experiment on auditory selective attention in reverberant room conditions, the stimuli are extended by a direction word. The stimuli are therefore longer and polysyllabic (3.4.2). Headphones as described in chapter 3.7.3 are used for all participants. Binaural synthesis is based on HRTFs of the dummy head, but headphones are equalized by individually measured HpTFs (compare chapter 3.6.3). The experiment takes place in the darkened hearing booth (compare chapter 3.5.2). Using RAVEN stimuli are adjusted to three different reverberation levels (compare chapter 3.8).

There are four independent variables in the present experiment (compare table 4.5). The variable of reverberation has three levels as described above (anechoic

vs. low reverberation vs. high reverberation) (different than in experiment III, this variable is a within subject variable in the present experiment).
Target's position (median vs. diagonal vs. frontal plane), attention switch (repetition vs. switch) and congruency (congruent vs. incongruent) are described in chapter 3.2. Dependent variables are reaction times and error rates.
The angular arrangement of target and distractor cannot be taken into account within the presented analysis due to limited data. In Oberem and colleagues [136] further information about the angular arrangement are presented.

"In total, 576 trials divided into six blocks of 96 trials each [...] [are] separated by short breaks (2 min and 5 min between the third and fourth block, respectively). The experimental blocks [...] [are] preceded by two training blocks. The first training block (10 trials) [...] [presents] the target's speech only to give the [...] [participant] the opportunity to get familiar with the input device. Another 40 trials [...] [are] presented in the second anechoic training's block, also including the distractor's speech as in the experimental blocks. The total duration of the experiment [...] [does] not exceed 70 min including an audiometry.
The three different reverberation conditions [...] [are] changed block-wise. Conditions [...] [are] assigned to block numbers according to a Latin Square Design. Within a block the location of the target speaker [...] [is] repeated or changed by the same chance. The location of the distracting speaker [...] [is] changed in every trial. Changes of speakers positions [...] [are] assigned randomly. Furthermore, trials [...] [are] counterbalanced over combinations of digits. The speakers of the stimuli [...] [are] assigned randomly"[136].

Table 4.5.: Experiment V: Independent variables and their levels.

Independent Variables	
Reverberation [R]	Anechoic Low Reverberation High Reverberation
Position of Target [TPOS]	Median Diagonal Frontal
Attention Switch [AS]	Repetition Switch
Congruency [C]	Congruent Incongruent

4.5.2. Results

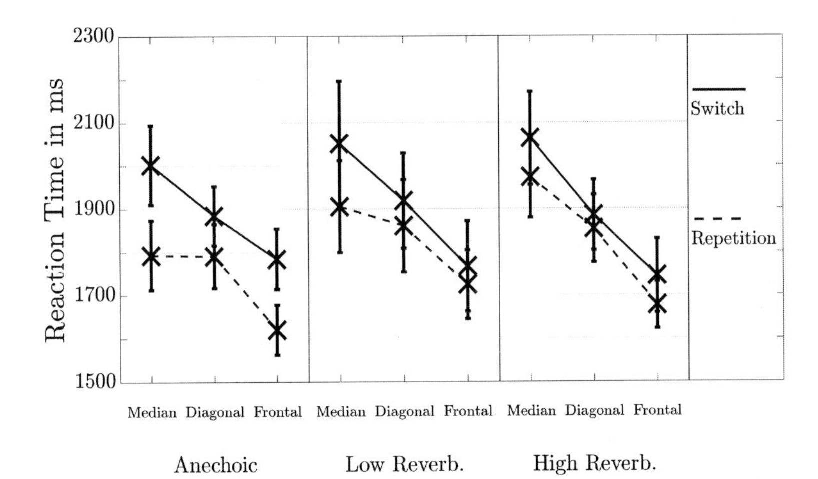

Figure 4.11.: Reaction times (in ms) as a function of reverberation, target's position and attention switch (R × TPOS × AS). Error bars indicate standard errors. Note that for the sake of clarity the variable of congruency is not visualized in this plot.

Reaction Times

In reaction times, there is no significant effect in reverberation time. There is a trend towards shorter reaction times in the anechoic condition (anechoic: 1812 ms vs. low reverb.: 1871 ms vs. high reverb.: 1866 ms) [R: $F < 1$].
The main effect of the target's position [$TPOS$: $F(1.51, 33.13) = 32.96$, $p < .001$, $\eta_p^2 = .60$] is significant. A post-hoc test reveals that reaction times are significantly longest for trials where the target is positioned in the median plane and significantly smallest for trials where the target is positioned in the frontal plane (Median: 1964 ms vs. Diagonal: 1865 ms vs. Frontal: 1719 ms).
There is a significant main effect of the attention switch, indicating longer reaction times for switches than for repetitions (1900 ms vs. 1799 ms) [AS: $F(1, 22) = 17.11$, $p < .001$, $\eta_p^2 = .44$] (compare figure 4.11). The switch costs amount on average to 101 ms.
The main effect of congruency is not significant [C: $F(1, 22) = 1.38$, $p = .18$, $\eta_p^2 = .08$]. A non-significant trend towards longer reaction times for incongruent trials than for congruent trials (incongruent: 1885 ms vs. congruent: 1814 ms) is

observed.
The ANOVA yields a significant interaction of reverberation time and attention switch [$R \times AS$: $F(2,44) = 3.45$, $p = .04$, $\eta_p^2 = .12$], indicating greater switch costs for anechoic conditions than for reverberant conditions (anechoic: 156 ms vs. low reverb.: 82 ms vs. high reverb.: 63 ms). A post-hoc test shows a significant difference between repetitions of the anechoic and the high reverberation condition ($p < .05$).

Error Rates

In error rates, there is a significant main effect of reverberation time [R: $F(2,44) = 3.94$, $p = .03$, $\eta_p^2 = .15$], indicating significant smaller error rates in anechoic conditions than in highly reverberating conditions (anechoic: 12.4 % vs. low reverb.: 13.4 % vs. high reverb.: 14.8 %).
The main effect of the target's position is significant [$TPOS$: $F(2,44) = 23.87, p < .001$, $\eta_p^2 = .52$]. Error rates are significantly highest for trials where the target is positioned in the median plane compared to the frontal plane (Median: 15.9 % vs. Diagonal: 14.9 % vs. Frontal: 9.8 %).
No significant attention switch effect is found in error rates [AS: $F < 1$].
The ANOVA yields a significant main effect of congruency [C: $F(1,22) = 231.76, p < .001$, $\eta_p^2 = .91$], indicating higher error rates for incongruent trials than for congruent trials (incongruent: 23.4 % vs. congruent: 3.7 %).
There is a significant interaction of reverberation time and congruency [$R \times C$: $F(2,44) = 5.38, p < .01$, $\eta_p^2 = .20$]. While congruent trials yield nearly the same error rates for all tested reverberation times, incongruent trials yield higher error rates with increasing reverberation time (congruency effect: anechoic: 17.8 % vs. low reverb.: 19.2 % vs. high reverb.: 21.3 %). Post-hoc tests indicate a significant difference in incongruent trials between anechoic and high reverberation conditions.
Furthermore, the interaction of the target's position and the congruency effect turns out to be significant [$TPOS \times C$: $F(2,44) = 27.74, p < .001$, $\eta_p^2 = .56$]. The congruency effect is greatest in median plane and smallest in frontal plane (difference in error rates between congruent and incongruent trials: median: 26.0 % vs. diagonal: 20.4 % vs. frontal: 12.9 %). Error rates from incongruent trials differ significantly between these where the target is positioned in frontal plane and those where it is positioned in median or diagonal plane.

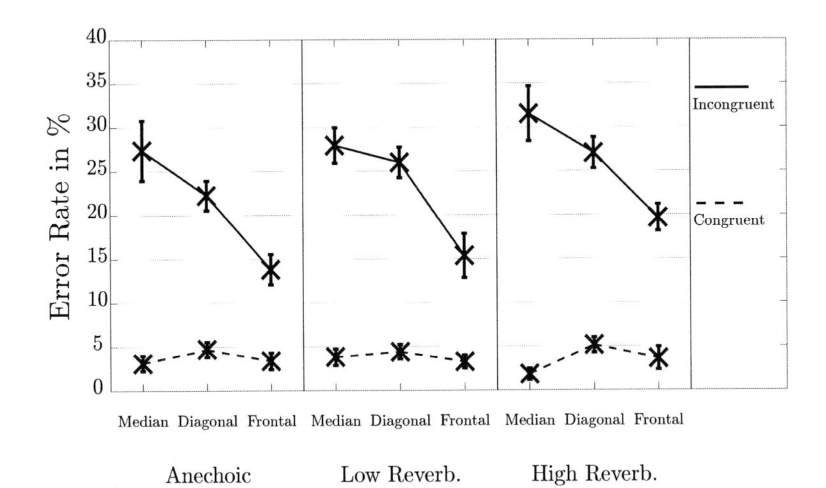

Figure 4.12.: Error rates (in %) as a function of reverberation, target's position and congruency (R × TPOS × C). Error bars indicate standard errors. Note that for the sake of clarity the variable of attention switch is not visualized in this plot.

4.5.3. Discussion

"In line with Kidd and colleagues [69] as well as Darwin and Hukin [29], reverberation significantly [...] [degrades] the performance in error rates in the range of 3.3 % between anechoic and high reverberation conditions. In addition to error rates and different to the cited investigations, reaction times [...] [are] also observed, where only a non-significant trend towards longer reaction times for higher reverberant energy [...] [is] found. However, the most important finding in this study [...] [is] the interaction between reverberation and the attention switch in reaction times. In accordance with previous findings [74, 85, 88, 129, 44] the attention switch [...] [is] significant in reaction times amounting to about 100 ms switch costs. These switch costs [...] [are] strongly depended on the reverberation time yielding to a maximum difference of 93 ms between the anechoic and the high reverberation condition (switch $\text{cost}_{\text{anechoic}} = 156$ ms and switch $\text{cost}_{\text{high reverberation}} = 63$ ms). This effect of decreasing switch costs for increasing reverberation [...] [is] based on the increasing reaction times for repetition trials for increasing reverberation. While under anechoic condition reaction times amount on average to 1734 ms, they [...] [are] up to 100 ms greater under high reverberation condition. Reaction times for switch trials [...] [do] not significantly

differ between reverberation conditions (max. difference: 23 ms). Consequently, intentionally switching auditory selective attention [...] [is] such a great demand by itself that additional reverberant energy [...] [does] not make the task more difficult regarding reaction times.
Independent from switch or repetition trials a significant interaction of reverberation and congruency [...] [is] observed in error rates. The congruence effect [...] [is] mainly reflected in error rates amounting to about 20 % difference in error rates between congruent and incongruent trials [...]. The interaction with reverberant energy [...] [shows] how the task of focusing on and processing the target speaker while ignoring the speech of the distracting speaker [...] [is] influenced by reverberation. Up to 5 % difference in error rates between anechoic and high reverberation condition for incongruent trials [...] [are] observed while error rates for congruent trials [...] [are] not affected by the reverberant energy. The congruence effect [...] [can] be taken as an implicit performance measure of attending to task relevant information and filtering out the irrelevant information [74] and hence, the conclusion [...] [is] drawn that reverberation significantly affected attending and filtering out information of target and interferer"[136].
"Reverberation has a detrimental effect on reaction times when maintaining attention to one source at a constant spatial location, however, intentionally switching the attention to a sound source at a different spatial location requires per se more attention and is more difficult that additional reverberant energy does not have any impact. Furthermore, the human ability to ignore or rather not to process the content of a distracting source is significantly influenced by reverberation"[136].
As a next step towards realistic auditory listening scenes the binaural reproduction method is converted from static to dynamic reproduction.

4.6. From Static to Dynamic – Experiment VI

Parts of this study are presented at the national conference on acoustics DAGA in 2017 [132]. Experimental data can be downloaded from the technical report [124].

In Experiment II (compare chapter 4.2) different reproduction methods are compared, resulting in (partly significant) differences between the reproduction method of real sources (A) and the individual binaural synthesis (B). It is assumed that these differences originate from the limitation of head movements and the consequential constraint in localization cues naturally offered by head movements (compare chapter 2.2.2). To illuminate this challenge and as a next step towards realistic environments, the dynamic binaural reproduction with consideration of head movements is compared to a static binaural reproduction.
To reproduce a realistic auditory scene it is often proven that a dynamic reproduction is more beneficial compared to a static reproduction. Pedersen and Minnaar [142], in line with others [66, 6, 23], report about a better localization accuracy, significantly less front-back-reversals as well as an impressing plausibility, when using a dynamic reproduction. In a localization task they presented either long stimuli (2 s) or short stimuli (2 s), resulting in much greater localization uncertainty when using short stimuli.
Wightman and Kistler [183] reassessed Wallach's experiments [177] on head movements in a modern manner. They support the thesis that head movements are not necessary to resolve ambiguities on median plane. However, a controlled movement of the source can also reduce front-back reversals. In the field of auditory scene analysis, Brinkmann and colleagues [19] examined authenticity applying static and dynamic binaural reproduction. In spite of the dynamic reproduction they report how the simulation is always clearly distinguishable from the reproduction using real sources. Nevertheless, for "non-critical source positions"[19] and speech stimuli the dynamic reproduction successful results are achieved.
On account of these findings, it is assumed that a dynamic reproduction should yield to more beneficial results compared to the static reproduction, which are compared in the present experiment [132].

4.6.1. Methods

A number of 23 paid, student participants aged between 19 and 35 years (mean age: 25.8 ± 4.8 years) take part in the experiment. They are equally divided in female and male participants. Listeners are screened to ensure that they have normal hearing (within 20 dB) for frequencies between 250 Hz and 10 kHz. All listeners can be considered as non-expert listeners since they have never participated in a listening test on auditory selective attention.

The binaural-listening paradigm as described in chapter 3.1.3 is used for this experiment and therefore stimuli presenting numbers combined with direction words are applied (compare chapter 3.4.2). Headphones as described in chapter 3.7.3 are used for all participants. Binaural synthesis is based on HRTFs of the dummy head, but headphones are equalized by individually measured HpTFs (compare chapter 3.6.3). The experiment takes place in the darkened hearing booth (compare chapter 3.5.2) equipped with an optical tracking system to monitor the participants' head movements. Using VA (compare chapter 3.7.4) HRTF filters are adjusted depending on the head orientation in real-time.

Table 4.6.: Experiment VI: Independent variables and their levels.

Independent Variables	
Reproduction Method [RM]	Static Quasi Static Real Dynamic
Position of Target [TPOS]	Median Diagonal Frontal
Attention Switch [AS]	Repetition Switch
Congruency [C]	Congruent Incongruent

The data of performed head movements are saved and interpreted. It is found that participants move rarely. Therefore, the reproduction method is split into three different levels, instead of intended categorization of "static" and "dynamic". The dynamic trials are split into the categories "quasi static" (containing 38 % of all data) and "real dynamic" (containing 12 % of all data). "Quasi static" comprises all trials that are reproduced dynamically, however, participants move less than 0.5 ° and therefore no HRTF filter change is initiated. Trials where HRTF filter changes are performed, are categorized into "real dynamic". Please note, that in several cases only one filter change is applied.

There are four independent variables in the present experiment (compare table 4.6). The variable reproduction method has three levels as described above (static vs. quasi static vs. dynamic).
Target's position (median vs. diagonal vs. frontal plane), attention switch (repetition vs. switch) and congruency (congruent vs. incongruent) are described in chapter 3.2. Dependent variables are reaction times and error rates.

In total 600 trials divided into four blocks of 150 trials each are separated by short breaks (5 min). Static and dynamic reproduction is presented block wise and are arranged in Latin square design. The experimental blocks are preceded by one training block of 50 trials. The total duration of the experiment does not exceed 60 min including the audiometry. Trials are counterbalanced over combinations of repetitions and switches, target's and distractor's postion.

4.6.2. Results

Reaction Times

In reaction times, the repeated measures ANOVA yields no significant main effect of reproduction method [RM: $F(2, 44) = 1.06$, $p = .36$, $\eta_p^2 = .05$].
The ANOVA reveals a significant main effect of target's position [$TPOS$: $F(2, 44) = 17.41$, $p < .001$, $\eta_p^2 = .44$] (compare figure 4.13). A significant post-hoc analysis determines significant differences ($p < .001$) in performance between the frontal plane and the other two planes, indicating longest reaction times for trials with target positioned on median plane and lowest for those on frontal plane (compare table A.18).
The main effect of attention switch on reaction time is significant [AS: $F(1, 22) = 17.41$, $p < .001$, $\eta_p^2 = .44$] and indicates longer reaction times for switches than for repetitions (compare figure 4.13). The switch costs amounted on average to 75 ms.
The ANOVA also yields no significant main effect of congruency [C: $F < 1$, $\eta_p^2 < .001$].
The interaction of reproduction method and target's position turns out to be significant [$RM \times TPOS$: $F(3.29, 72.43) = 3.64$, $p = .009$ $\eta_p^2 = .14$]. Reaction times significantly differ between the static and the real dynamic reproduction when

the target is positioned on median plane, indicating shorter reaction times for the dynamic reproduction. Furthermore, quasistatic and dynamic reproduction differ significantly on frontal plane, resulting in longer reaction times for the dynamic reproduction.

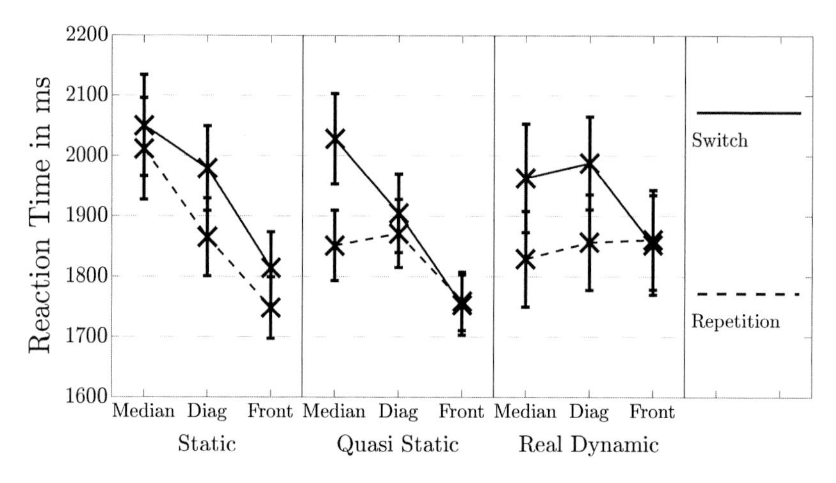

Figure 4.13.: Reaction times (in ms) as a function of reproduction method, target's position and attention switch (RM × TPOS × AS). Error bars indicate standard errors. Note that for the sake of clarity the variable of congruency is not visualized in this plot.

Error Rates

In error rates, the repeated measures ANOVA yields no significant main effect of the variable reproduction method [RM: $F(2,44) = 2.60$, $p = .09$, $\eta_p^2 = .11$].
The ANOVA reveals a significant main effect of target's position [$TPOS$: $F(2,44) = 17.14$, $p < .001$, $\eta_p^2 = .44$] (compare figure 4.14). A significant post-hoc analysis determines significant differences ($p < .001$) in performance between frontal plane and the other two planes, indicating highest error rates for trials with target positioned on median plane and lowest for those on frontal plane (compare table A.18).
The main effect of attention switch on reaction time is not significant [AS: $F < 1$, $\eta_p^2 = .04$].
The ANOVA yields a significant main effect of congruency [C: $F(1,22) = 86.42$, $p < .001$ $\eta_p^2 = .80$], indicating higher error rates for incongruent stimuli than for

congruent stimuli (23.5 % vs. 9.5 %).
The interaction of target's position and congruency turns out to be significant [$TPOS \times C$: $F(2,44) = 5.25$, $p < .009$ $\eta_p^2 = .19$]. Error rates for congruent trials are significantly different on median and frontal plane. For incongruent trials results differ significantly for the frontal plane compared to the other two planes.

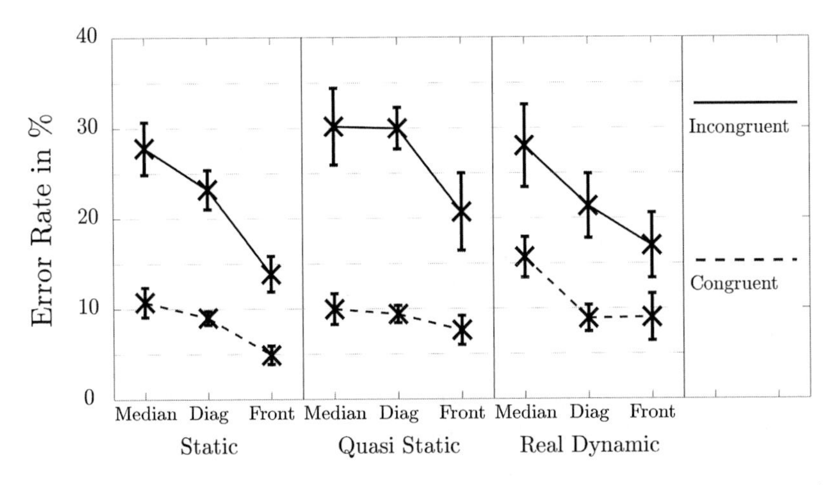

Figure 4.14.: Error rates (in %) as a function of reproduction method, target's position and congruency (RM × TPOS × C). Error bars indicate standard errors. Note that for the sake of clarity the variable of attention switch is not visualized in this plot.

4.6.3. Discussion

The present investigation examined how a static and a dynamic reproduction affected the intentional switching in auditory selective attention [132]. Results are contradictory: On the one hand, in reaction times, there is a significant improvement in median plane for the dynamic reproduction, possibly due to the frequently confirmed effect of diminishing front-back-confusions in dynamic reproductions [142]. On the other hand, the dynamic reproduction yields significantly longer reaction times in frontal plane than the quasi static reproduction. In error rates no differences between the static, the quasi static and the real dynamic reproduction can be found.
In a localization experiment, performed as a technical pretest, exactly the same dynamic reproduction is used [135, 148]. The main focus of this localization

experiment is placed on the direct comparison of static and dynamic reproduction with different resolutions in HRTF-data. Dynamic reproduction of any resolution applied turned out fundamental for a reduction of undesired front-back reversals and in-head localization, which is in line with cited localization experiments [142, 66, 6, 23]. Due to this technical pretest it is assured that the present reproduction system is not responsible for the missing effects between static and dynamic reproduction, especially relating to interactions with auditory attention switches and congruency.

The main finding of this experiment is that the observed head movements are very small. In preceding experiments (for example chapter 4.2) head movements are monitored and trials in which head movements exceeded 2° in rotation are excluded from the analysis. However, detailed trajectories of head movements are not saved or evaluated. Even though Blauert [11] reported on head movements that are consistently small (compare chapter 2.2.2), greater movement angles in an auditory scene using a task of switching attention are assumed.

When discussing the results of experiment II, it is assumed that differences between the reproduction methods A and B (real sources vs. individual binaural synthesis) is due to the static presentation of the binaural synthesis in reproduction method B. Since no significant differences in error rates between the dynamic and the static reproduction can be found, the difference in experiment II most probably originate from any other effect. As a next step to identify the reason for the significant difference the reproduction with real sources should be compared to a dynamic reproduction based on individual binaural synthesis. It should be taken into consideration that Brinkmann and colleagues [19], who examined authenticity of dynamic binaural synthesis, reported how participants are always able to identify the synthesis from real sources despite the dynamic reproduction. Therefore, it may happen that differences between the reproduction methods also appear with the present paradigm.

Based on the knowledge taken from the present experiment, it is observed that performed head movements are mostly smaller than 0.5° and assumed that this is also due to restrictions in the experimental setup. Participants look at the cue on a monitor straight ahead and know that missing this cue results in failing to answer correctly to the task. Participants are reminded before every experimental block, that reaction times and error rates should be both as low as possible. This also yields to a restricted time window of one trial. It is believed that the presentation of the cue is a constraint of the extended binaural-listening paradigm when examining auditory selective attention in realistic environments. For realistic scenes with moving participants and sources the presentation of the cue has to be reassessed.

4.7. Age-related Effects – Experiment VII

Parts of this study are presented at the national conference on acoustics DAGA in 2015 [130] and published in ACTA psychologica in 2017 [134]. Experimental data can be downloaded from the technical report [124].

"All reported effects with the binaural-paradigm on intentional attention switching [...] [are] found with young participants (18-35 years). As there is a trend towards an aging society, especially in western civilization, age-related effects have become of greater interest [105, 151]. There [...] [are] already several investigations on age-related effects in attention working with dichotic reproduction that report increased performance costs in older people [60, 71, 76, 75, 79, 77, 80, 78, 104, 108, 155, 174].
Age-related switch costs for guided attention switches [...] [are] examined by Tun and Lachman [170] and Lawo and Koch [86]. Tun and Lachman's results of larger switch costs for older participants [...] [are] in contrast to these of the dichotic listening investigation by Lawo and Koch [86] who use the described [dichotic-listening] paradigm [74] examining the ability to prepare for an upcoming auditory attention switch. Lawo and Koch [86] report that older participants respond significantly slower than younger participants. However, the attention switch costs [...] [do] not differ across age groups, confirming the idea of "general slowing", as confirmed in a meta-analysis by Wasylyshyn and colleagues [178], who [...] [do] not find age-related differences in task switching using visual tasks. There are several theories of cognitive aging assuming that the ability to inhibit irrelevant information declines with age [17, 55, 108]; therefore it [...] [can] be assumed that an increased congruency effect in older people [...] [is] likely. However, Lawo and Koch's [86] findings [...] [do] not correspond to predictions from the inhibitory deficit theory, since they neither [...] [find] an increased congruency effect nor increased switch costs for older participants.
Age-related effects have also been analyzed with binaural-listening-setups. The effect that younger participants outperform older participants [...] [is] often observed in investigations that focused on tasks of perceiving competing speech [36, 59, 63, 93, 100, 166, 172, 171].
Multiple source possibilities in a binaural-listening-setup build a more complex scene than a dichotic presentation of stimuli. To successfully focus on the stimulus of the target speaker in a binaural setup, the ability to localize different sound sources [...] [is] necessary. The age-related effect in localization tasks [...] [is] for example analyzed by Abel [and colleagues] [1]. They focus on sound localization on the horizontal plane for participants aged between 10 and 81. Performance of older adults [...] [decreases], especially in front-back-confusions and on the right

side of space. The deterioration of accuracy and precision by older individuals [...] [is] also found by Dobreva [and colleagues] [34].
Age-related effects in involuntary switches in binaural-listening-setups [...] [are] examined by Singh [and colleagues] [166]. Using the Coordinate Response Measure Corpus [15], participants [...] [are] asked to repeat the color and number word preceded by a fixed call-sign and consequently correct word-identification scores [...] [are] measured. Participants [...] [are] provided with advance information about the probability of the sentence being presented from one out of three possible frontal positions. No age-related differences in switching attention from one location to another [...] [are] found. However, in a more complex task where the participant's attention [...] [is] intentionally misdirected and the participant [...] [is] therefore required to perform multiple switches of attention, age-related deficits [...] [are] reported.
In the present investigation [...] the binaural reproduction of stimuli and the analysis of age-related effects in a task of instructed attention switches [is combined]. Based on the findings of the previous binaural-listening experiment, two groups of different age [...] [are] tested in the binaural-listening-setup. In general, [...] [it is expected that] age-related effects in intentionally switching auditory selective attention [...] [are]comparable to these of the previous dichotic investigation [86]. Essential differences between this investigation and the investigation by Lawo and Koch [86] [...] [are] the reproduction method (dichotic vs. binaural) and the cue criterion (gender vs. location). Regarding the location of the target speaker [...] differences in reaction times and error rates [are expected]. However, there [...] [are] no firm expectations about the interaction of age group with attention switches and congruency"[134].

4.7.1. Methods

"A number of 20 paid, student participants aged between 20 and 31 years (mean age: 24.5 ± 3.1 years) as well as 20 paid senior citizen participants aged between 58 and 74 (mean age: 67.8 ± 3.3 years) [...] [take] part in the experiment. They [...] [are] equally divided in female and male [...] [participants]. All listeners [...] [can] be considered as non-expert listeners since they [...] [have] never participated in a listening test on auditory selective attention. All listeners [...] [are] screened by an ascending-pure-tone-audiometry procedure for frequencies between 125 Hz and 8 kHz. All younger participants [...] [have] normal hearing (within 25 dBHL defined as no impairment by the WHO [138], no greater between-ear-difference than 8 dB in all tested frequencies). Older participants suffer from a slight hearing loss in higher frequencies, but none of them [...] [is] provided with any hearing aid. Composite audiograms for both participant groups are shown in Figure 4.15. Based on the categorization of hearing impairment by the WHO [138], one older participant [...] [belongs] to the category of slight impairment (25.89 dBHL) and all others [...] [do] not show any impairment. The data of the older participant with the slight impairment [...] [is] not taken out of the analysis since results [...] [are] above-average"[134].

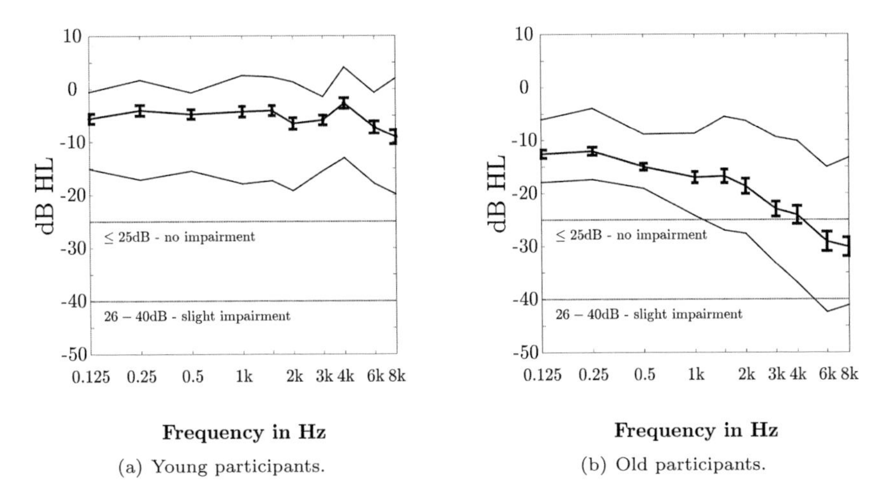

(a) Young participants.

(b) Old participants.

Figure 4.15.: Audiogram of young participants (a) and older participants (b). The thick solid line indicates mean values with standard errors per measured frequency. Upper and lower solid lines represent maximum and minimum values of all participants. Categorization of hearing impairment by the WHO is represented by dotted lines.

The former binaural-listening paradigm as described in chapter 3.1.2 is used for this experiment on age effects. Headphones as described in chapter 3.7.3 are used for all participants. Binaural synthesis is based on HRTFs of the dummy head, but headphones are equalized by individually measured HpTFs (compare chapter 3.6.3). The experiment takes place in the darkened hearing booth (compare chapter 3.5.2) [130].

There are four independent variables in the present experiment (compare table 4.7). The between-subject variable of age has two levels (young vs. old).
Target's position (median vs. diagonal vs. frontal plane), attention switch (repetition vs. switch) and congruency (congruent vs. incongruent) are described in chapter 3.2. Dependent variables are reaction times and error rates.

"In total, 432 trials divided into three blocks of 144 trials each [...] [are] separated by short breaks (5 min). One training block of 50 trials [...] [precedes] the experimental blocks. The total duration of the experiment [...] [does] not exceed 50 min, including the audiometry. Trials [...] [are] counterbalanced over combinations of number words with randomly assigned speakers."[134]

Table 4.7.: Experiment VII: Independent variables and their levels.

Independent Variables	
Age [A] (between-subject)	Young Old
Position of Target [TPOS]	Median Diagonal Frontal
Attention Switch [AS]	Repetition Switch
Congruency [C]	Congruent Incongruent

4.7.2. Results

Main Effects – Reaction Times

For reaction times, all main effects turn out to be significant (compare table A.19).
The repeated measures ANOVA yields a significant main effect of the between-subject variable age [A: $F(1,38) = 23.52$, $p < .001$, $\eta_p^2 = .38$], indicating longer reaction times (1731 ms) for old participants compared to young participants (1152 ms) (compare figure 4.16 and table A.21).
The Huynh-Feldt corrected ANOVA reveals a significant main effect of target's position [$TPOS$: $F(1.34, 51.07) = 30.93$, $p < .001$, $\eta_p^2 = .45$]. A significant post-hoc analysis determines significant differences ($p < .01$) in performance between all planes, indicating longest reaction times for trials with target positioned on median plane ($\sim$ 1540 ms) and lowest for those on frontal plane ($\sim$ 1317 ms) (compare table A.21).
The main effect of attention switch on reaction time is significant [AS: $F(1,38) = 21.43$, $p < .001$, $\eta_p^2 = .36$] and indicates a longer reaction time for switches than

for repetitions. The switch costs amount on average to 90 ms.
The ANOVA also yields a significant main effect of congruency [C: $F(1, 38) = 32.44$, $p < .001$ $\eta_p^2 = .46$], indicating longer reaction times for incongruent stimuli (1513 ms) than for congruent stimuli (1369 ms).

Significant Interactions – Reaction Times

The target's position interacts with the variable age [$TPOS \times A$: $F(1.34, 51.07) = 5.83$, $p = .01$, $\eta_p^2 = .13$]. While for old participants the reaction times differ significantly between all planes, for young participants reaction times are only significantly different on frontal compared to diagonal plane, respectively. Standard errors on median plane are very high (1201 ± 105 ms), which is why no significant difference occured.
The congruency interacts with the between-subject variable age [$C \times A$: $F(1, 38) = 12.86$, $p = .001$, $\eta_p^2 = .25$]. The congruency effect is only significant for old participants in reaction times (congruency effect: old:235 ms, young:53 ms) (compare figure 4.16).
The target's position interacts with the attention switch [$TPOS \times AS$: $F(1.66, 62.99) = 4.41$, $p = .02$, $\eta_p^2 = .10$], indicating greatest switch costs on median plane and lowest on frontal plane (Switch costs: median 134 ms, diagonal 54 ms, frontal 84 ms.
The target's position interacts also with the variable congruency [$TPOS \times C$: $F(2, 76) = 24.22$, $p < .001$, $\eta_p^2 = .39$], indicating a significant congruency effect on median and diagonal plane (Congruency effect: median 238 ms, diagonal 156 ms, frontal 39 ms.
The three-way interaction with age turns out to be significant [$TPOS \times C \times A$: $F(2, 76) = 11.60$, $p < .001$, $\eta_p^2 = .23$], showing how the significant congruency effect is true for all planes for old participants and never true for younger participants in reaction times (compare figure 4.16).
Please find further interactions in table A.19.

Main Effects – Error Rates

For error rates, all main effects turn out to be significant (compare table A.20). The repeated measures ANOVA yields a significant main effect of the between-subject variable age [A: $F(1, 38) = 21.19$, $p < .001$, $\eta_p^2 = .36$], indicating larger error rates (16.1 %) for old participants compared to young participants (8.2 %) (compare figure 4.17 and table A.21).
The ANOVA reveals a significant main effect of target's position [$TPOS$:

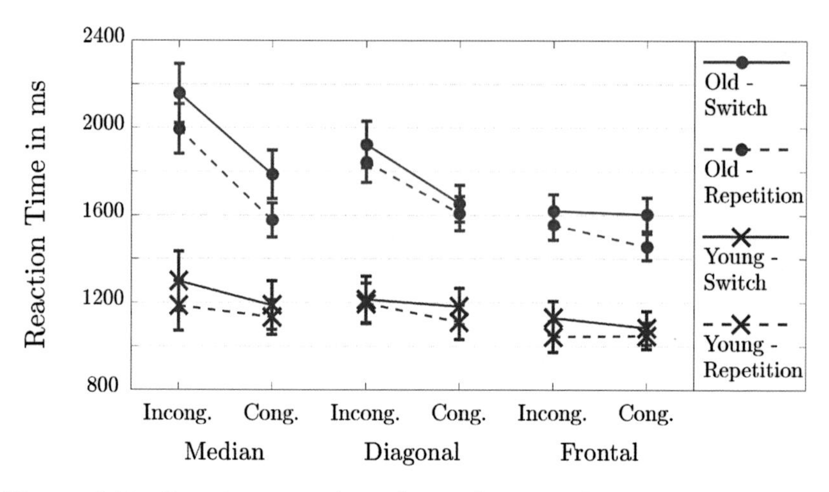

Figure 4.16.: Reaction times (in ms) as a function of age, target's position, attention switch and congruency (A × TPOS × AS × C). Error bars indicate standard errors.

$F(1.76, 66.87) = 55.42$, $p < .001$, $\eta_p^2 = .59$]. A significant post-hoc analysis determines significant differences ($p < .01$) in performance between all planes, indicating highest error rates for trials with target positioned on median plane (~ 15.8 %) and lowest for those on frontal plane (~ 7.9 %) (compare table A.21). The main effect of attention switch on reaction time is significant [AS: $F(1, 38) = 4.86$, $p = .03$, $\eta_p^2 = .11$] and indicates higher error rates for switches than for repetitions. The switch costs amounted on average to 0.9 %.
The ANOVA also yields a significant main effect of congruency [C: $F(1, 38) = 116.40$, $p < .001$ $\eta_p^2 = .75$], indicating higher error rates for incongruent stimuli (20.2 %) than for congruent stimuli (4.1 %).

Significant Interactions – Error Rates

The target's position interacts with the variable age [$TPOS \times A$: $F(1.76, 66.87) = 7.38$, $p = .002$, $\eta_p^2 = .16$]. While for old participants the error rates differ significantly between all planes, for young participants error rates are only significantly different on frontal plane compared to the other two planes, respectively.
The congruency interacts with the between-subject variable age [$C \times A$: $F(1, 38) = 7.34$, $p = .01$, $\eta_p^2 = .16$]. The congruency effect is significanly greater for old participants in error rates compared to young participants (congruency effect:

old:20.1 %, young:12.0 %) (compare figure 4.17).
The target's position interacts with the variable congruency [$TPOS \times C$: $F(2, 76) = 27.19$, $p < .001$, $\eta_p^2 = .42$], indicating the significantly greatest congruency effect on median plane (Congruency effect: median: 21.8 %, diagonal: 17.0 %, frontal: 9.4 %).

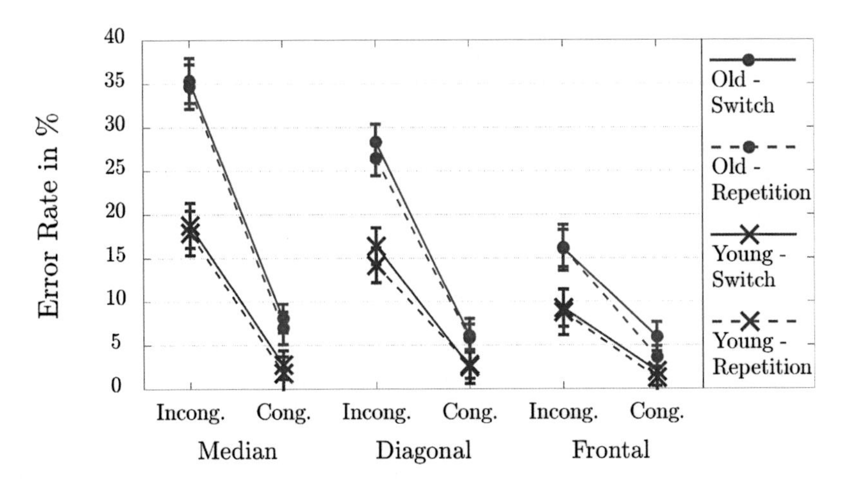

Figure 4.17.: Error rates (in %) as a function of age, target's position, attention switch and congruency (A × TPOS × AS × C). Error bars indicate standard errors.

4.7.3. Discussion

"Reaction times and error rates of young, normal-hearing and older, normal-hearing to slightly hearing-impaired participants [...] [are compared]. There [...] [is] statistical evidence that absolute values of reaction times and error rates differ between age groups in this investigation. These expected results agree with the previous dichotic investigation on age-related effects in intentionally switching auditory selective attention [86]. Increased reaction times and error rates for older participants [...] [are] often found in dichotic and binaural investigations on attention [1, 34, 36, 49, 59, 63, 75, 79, 93, 100, 166, 172, 174].
Along with the binaural extension, more spatial positions for the location of target and distractor than in a dichotic listening setup [...] [are] selected and their influence [...] [is] analyzed. While target locations to the left and to the right [...] [are] comparably the easiest positions for young and old adults to focus on,

target locations in the median plane (front and back) [...] [are] the most difficult positions to attend for young participants and even significantly more for older adults. In their localization experiment with participants of different age groups, Abel and colleagues [1] also found greater difficulties for sources positioned in front and back than for lateral source positions. The localization performance from younger (10-39 years) to older (60-81 years) participants dropped about 8 % for lateral positions and about 12.5 % for positions on or close to the median plane. Regardless of age-related effects, deteriorated performance in the median plane compared to other positions on the horizontal plane [...] [is] often observed [116], especially when binaural stimuli [...] [are] based on non-individual HRTFs as in the present study. It can be summarized that it [...] [is] most difficult to focus attention on sources positioned in the median plane and an age-related effect on the horizontal plane compared to other source positions [...] [is] found. The significant effect of endogenous attention switch, indicating that participants respond faster when the target's direction [...] [is] repeated, [...] [can] be observed with both age groups as well as in previous investigations of the authors [74, 86, 129] using dichotic and binaural listening. Switch costs, which provide an explicit measure of how well instructions to switch attention [...] [can] be followed, [...] [do] not differ significantly from those of the previous binaural investigation [129]. To successfully perform the task of attention switching the inhibition of competing perceptual filter settings [...] [may] be important. That there is an age-related decline of the ability to inhibit irrelevant information has been predicted in several theories [17, 55]. In this investigation, however, the auditory switch costs in older participants [...] [are] similar compared to those of young participants, suggesting no age-related differences in attention switching. This confirms previous findings using a simpler dichotic-listening set-up [86]. Assuming that inhibitory processes contribute to auditory switch costs, the results of this investigation deviate from the inhibitory deficit theory. As reported by Singh and colleagues [166], age-related deficits in word-identification scores [...] [do] also not occur in simple tasks of exogenous attention switches. However, significant age-related attentional deficits [...] [are] detected in more complicated tasks of multiple attention switches. The findings in error rates agree with the present study. Even though the present result of non-significant age-related differences on switch costs in reaction times (in addition to error rates) represent a null effect, it nevertheless provided additional evidence that older participants perform similar to young participants in tasks of intentional attention switches.

The examination of the congruency of number words of target and distractor [...] [offers] the opportunity to analyze the ability of younger and older participants to focus their attention on one speaker and simultaneously ignore the distracting speaker. The greatest difference between the dichotic investigation [86] and the present binaural investigation [...] [can] be found in the interaction of congruency and the age-related effect when looking at the non-logarithmic scores. Both investigations show a significant effect in reaction times and error rates in the main effect of congruency, indicating a worse performance for incongruent trials. Indeed, the present investigation [...] [shows] a significant interaction in reaction times, indicating that older adults perform comparatively worse when stimuli [...] [are] incongruent which [...] [can] not be found in the dichotic investigation. The difference between congruent and incongruent trials in reaction time [...] [is] proportionally three times greater for older participants than for young adults. It [...] [can] be assumed that older people have more difficulties to ignore a second speaker than younger adults in a binaural-listening-setup. Thus, it [...] [appears] that the current results [...] [are] in line with the hypothesis of older adults having a deficit in inhibitory processes [17, 55]. Considering that there [...] [is] no age-related effect in attention switch costs, it [...] [may] be assumed that ignoring concurrent speech depends on inhibition to a much higher degree than switching attention.

In contrast to present findings [...] [is] an examination of age-related inhibition of irrelevant speech by Li and colleagues [93]. In a binaural setup with two source positions, Li and colleagues test the ability to inhibit the masker's speech with older and young participants in a shadowing task using meaningless sentences. Since older adults do not have more difficulties inhibiting the irrelevant, informational masker in this examination, Li and colleagues' results oppose the inhibitory deficit theory. Differences [...] [may] arise from the different complexities of the binaural setups. The binaural setup used by Li and colleagues [...] [is more] simple, compared to the binaural setup including eight sources around the listener used in the present study.

A possible explanation for this difference between the dichotic and the present investigation [...] [may] also be found in the source setups. The congruency effect [...] [interacts] with the effect of the target's spatial position, indicating that the congruency effect [...] [is] greatest for the target positioned on the median plane and smallest for the target positioned to the right or left of the participant. The interaction with age [...] [shows] significantly [...] [longer] reaction time differences between source positions for older than for younger participants. The effect that older participants [...] [are] significantly more distracted by the opposing speaker for target positions in the median plane than for positions in the diagonal plane or to the sides [...] [can] not be explained completely. It [...] [may] be assumed that the applicability of the inhibitory deficit theory is confined to dichotic or very simple spatial listening-test setups which [...] [are] mainly used to support this theory [17, 55, 86]. Evidence [...] [may] be given by the congruency effect

which [...] [is] least pronounced on the frontal plane (left and right) which is comparable to a very simple spatial setup.
As shown in Oberem and colleagues [129], young participants show significantly worse results in ignoring the distractor's speech as to be seen in the congruency effect when binaural stimuli [...] [are] non-individual. Therefore, it [...] [may] be assumed that older people also have greater difficulties in performing the task of attending a target-speaker while ignoring the opposing speaker with non-individual binaural stimuli. Possibly, older participants even suffer more from the loss of individual binaural information. Individual binaural information [...] [is] especially important for sources located in the median plane or for competing sources on one cone of confusion [11]. Effects between target positions on the median plane and other positions around the listener reinforce this thesis. Furthermore, the non-individual HRTFs [...] [are] measured with an artificial head, built from the image of a young person [111] and therefore, the data might have a better matching for young participants, since size and shape [...] [are] age-related [139]"[134].
Since an anechoic reproduction of a multi-talker scene fails to represent a realistic communication scenario, age effects are examined in reverberant environments in the subsequent chapter.

4.8. Age-related Effects under Reverberation – Experiment VIII

Parts of this study are presented at the national conference on acoustics DAGA in 2019 [137]. Experimental data can be downloaded from the technical report [124].

Experiment V (compare chapter 4.5) compares different reverberation levels within the paradigm of intentional switching of auditory selective attention employing young participants. To analyze a possible age-related effect in correlation with reverberation time, elderly participants (61-75 years) are tested and compared to the data of Experiment V in the present chapter.
Experiment V is related to the study by Ruggles and Shinn-Cunningham [152] (compare also chapter 4.3). Ruggles and Shinn-Cunningham tested young participants (18-35 years) and middle-agers (36-55 years) in three different reverberation settings. They do not find any age-related effect regarding reverberation time. Marrone and colleagues [101] employed participants within the same age-range (60-80 years) as in the present study. They analyze the interaction between hearing loss, reverberation, and age in a binaural set of multiple talkers. It is reported how correlations between spatial release and age are weak and performance differences can not be unambiguously attributed to the participant's age independent of hearing status. These findings are also consistent with the findings of Li and colleagues [93].
Significant age-related effects concerning reverberation could also not be found by Helfer and Wilber as well as Helfer and colleagues [58, 56]. They examined the accuracy of consonant identification in monaural and binaural presentation in noisy and reverberating listening conditions testing younger normal-hearing adults and older adults with little hearing loss. Most interesting finding is a negative correlation of age and performance in the reverberating noise condition. Based on the previous findings, it is assumed that no age-related effect regarding reverberation will be found.

4.8.1. Methods

A number 22 paid senior citizen participants aged between 61 and 75 (mean age: 66.8 ± 4.8 years) take part in the experiment. Data of 24 paid, student participants aged between 20 and 34 years (mean age: 23.9 ± 3.4 years) is taken from Experiment V to compare age groups. They are equally divided in female and male participants. All listeners can be considered as non-expert listeners since they have never participated in a listening test on auditory selective attention. All listeners are screened by an ascending-pure-tone-audiometry procedure for frequencies between 125 Hz and 8 kHz. All younger participants have normal hearing (within 25 dBHL defined as no impairment by the WHO [138], no greater between-ear-difference than 10 dB in all tested frequencies). Older participants suffered from a slight hearing loss in higher frequencies, but none of them was provided with any hearing aid. Composite audiograms for both participant groups are shown in Figure 4.18.

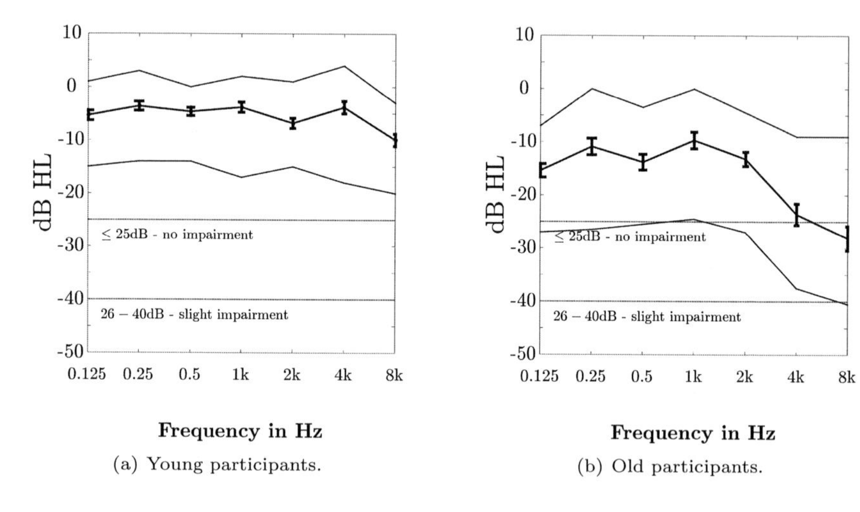

(a) Young participants.

(b) Old participants.

Figure 4.18.: Audiogram of young participants (a) and older participants (b). The thick solid line indicates mean values with standard errors per measured frequency. Upper and lower solid lines represent maximum and minimum values of all participants. Categorization of hearing impairment by the WHO is represented by dotted lines.

The extended binaural-listening paradigm as described in chapter 3.1.3 is used for this experiment, as an extension of Experiment V. The long and polysyllabic stimuli are compounded of a digit and a direction word (3.4.2). Headphones as described in chapter 3.7.3 are used for all participants. Binaural synthesis is based on HRTFs of the dummy head, but headphones are equalized by individually measured HpTFs (compare chapter 3.6.3). The experiment takes place in the darkened hearing booth (compare chapter 3.5.2). Using RAVEN stimuli are adjusted to three different reverberation levels (compare chapter 3.8).

There are five independent variables in the present experiment (compare table 4.8). The between-subject variable of age has two levels (young vs. old). The

variable of reverberation has three levels (anechoic vs. low reverberation vs. high reverberation) (also compare experiment V, chapter 4.5). Target's position (median vs. diagonal vs. frontal plane) is described in chapter 3.2.3.
There is not enough data collected to analyze all five variables at the same time. (For reaction times error trials and subsequent trials are deleted, mean error rates of 20 %, result in a rejection of up to half of all collected data.) Therefore, attention switch (repetition vs. switch) is only included in the analysis of reaction times and congruency (congruent vs. incongruent) in the analysis of error rates, respectively. Dependent variables are reaction times and error rates.

In total, 720 trials divided into six blocks of 120 trials each are separated by short breaks (2 min and 5 min between the third and fourth block, respectively). "The experimental blocks [...] [are] preceded by two training blocks. The first training block (10 trials) [...] [presents] the target's speech only to give the [...] [participant] the opportunity to get familiar with the input device. Another 40 trials [...] [are] presented in the second anechoic training's block, also including the distractor's speech as in the experimental blocks. The total duration of the experiment [...] [does] not exceed 70 min including an audiometry.
The three different reverberation conditions [...] [are] changed block-wise. Conditions [...] [are] assigned to block numbers according to a Latin Square Design. Within a block the location of the target speaker [...] [is] repeated or changed by the same chance. The location of the distracting speaker [...] [is] changed in every trial. Changes of speakers positions [...] [are] assigned randomly. Furthermore, trials [...] [are] counterbalanced over combinations of digits. The speakers of the stimuli [...] [are] assigned randomly"[136].

4.8.2. Results

Reaction Times

The repeated measures ANOVA yields a significant main effect of the between-subject variable age [A: $F(1,44) = 14.83$, $p < .001$, $\eta_p^2 = .25$], indicating longer reaction times (2255 ms) for old participants compared to young participants (1836 ms) (compare table A.22, figure 4.19 and table A.24).
The main effect of reverberation is not significant [R: $F < 1$, $\eta_p^2 = .01$].
The Huynh-Feldt corrected ANOVA reveals a significant main effect of target's position [$TPOS$: $F(1.35, 59.22) = 21.30$, $p < .001$, $\eta_p^2 = .33$]. A significant post-hoc analysis determines significant differences ($p < .01$) in performance between all planes, indicating longest reaction times for trials with target positioned on median plane (2132 ms) and lowest for those on frontal plane (1947 ms) (compare

Table 4.8.: Experiment VIII: Independent variables and their levels.

Independent Variables	
Age [A] (between-subject)	Young Old
Reverberation [R]	Anechoic Low Reverberation High Reverberation
Position of Target [TPOS]	Median Diagonal Frontal
Attention Switch [AS]	Repetition Switch
Congruency [C]	Congruent Incongruent

table A.24).
The main effect of attention switch on reaction time is significant [AS: $F(1, 44) = 23.54$, $p < .001$, $\eta_p^2 = .35$] and indicates longer reaction times for switches than for repetitions. The switch costs amount on average to 77 ms.

The target's position interacts with the attention switch [$TPOS \times AS$: $F(2, 88) = 3.25$, $p = .04$, $\eta_p^2 = .07$]. Switch costs are only significantly different on median plane (96 ms) and diagonal plane (97 ms). On frontal plane switch costs only amount to 36 ms.
The three-way interaction with age turns out to be also significant [$TPOS \times AS \times A$: $F(2, 88) = 3.86$, $p = .03$, $\eta_p^2 = .08$], showing how the significant attention switch effect is true for all planes for young participants and only true on diagonal plane for older participants.

Error Rates

For error rates, all main effects turn out to be significant (compare table A.23). The repeated measures ANOVA yields a significant main effect of the between-subject variable age [A: $F(1, 44) = 7.31$, $p = .01$, $\eta_p^2 = .14$], indicating larger error rates (20.9 %) for old participants compared to young participants (15.0 %) (compare figure 4.20 and table A.24).

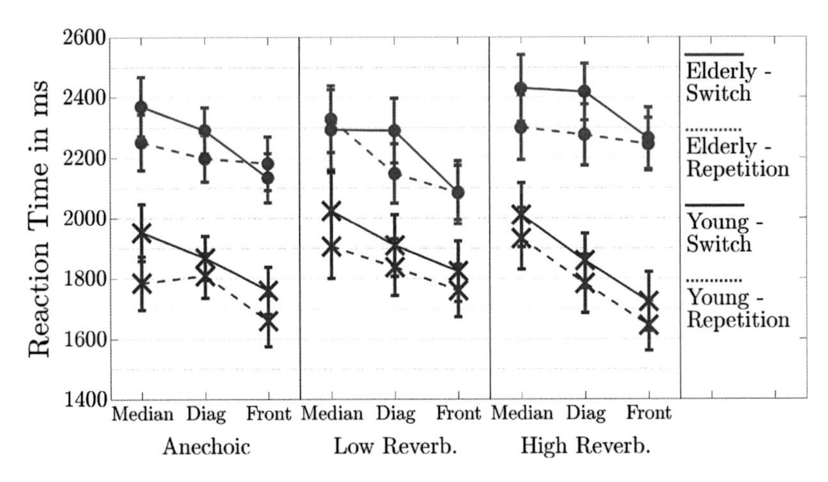

Figure 4.19.: Reaction times (in ms) as a function of age, reverberation, target's position and attention switch (A × R × TPOS × AS). Error bars indicate standard errors.

The main effect of reverberation is significant [R: $F(2, 88) = 6.57$, $p = .002$, $\eta_p^2 = .13$]. Post-hoc tests reveal a significant difference in error rates between the high reverberation and the other two reverberation conditions.
The ANOVA reveals a significant main effect of target's position [$TPOS$: $F(1.51, 66.59) = 52.77$, $p < .001$, $\eta_p^2 = .55$]. A significant post-hoc analysis determines significant differences in performance between all planes, indicating highest error rates for trials with target positioned on median plane (21.6 %) and lowest for those on frontal plane (13.4 %) (compare table A.24).
The ANOVA also yields a significant main effect of congruency [C: $F(1, 44) = 368.37$, $p < .001$ $\eta_p^2 = .89$], indicating higher error rates for incongruent stimuli (31.0 %) than for congruent stimuli (4.9 %).

The target's position interacts with the variable age [$TPOS \times A$: $F(1.51, 66.59) = 6.36$, $p = .006$, $\eta_p^2 = .13$]. The age-related effect is only true on median plane and not significant for the other two planes.
The congruency interacts with the between-subject variable age [$C \times A$: $F(1, 44) = 25.58$, $p < .001$, $\eta_p^2 = .37$]. In congruent trials, error rates of older and younger participants do not significantly differ and vary only about 1 %. In contrast to that, results of older and younger participants differ significantly, resulting in a difference of 12.8 % (compare figure 4.20).
The target's position interacts with the variable congruency [$TPOS \times C$:

$F(1.70, 74.81) = 58.63$, $p < .001$, $\eta_p^2 = .57$], indicating the greatest congruency effect on median plane (Congruency effect: median 34.9 %, diagonal 25.8 %, frontal 17.8 %.
The three-way interaction with age turns out to be also significant [$TPOS \times C \times A$: $F(1.70, 74.81) = 6.08$, $p = .006$, $\eta_p^2 = .12$], showing how the congruency effect increases with the complexity of planes (frontal, diagonal, median) for young and old participants. However, this increase of the congruency effect is more pronounced for older participants (compare figure 4.20).

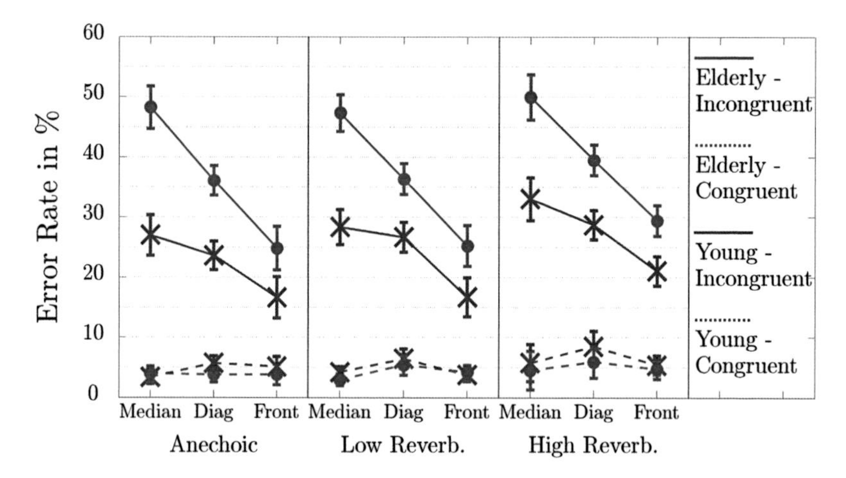

Figure 4.20.: Error rates (in %) as a function of age, reverberation, target's position and congruency (A × R × TPOS × C). Error bars indicate standard errors.

4.8.3. Discussion

Results of the present study are in compliance with those of experiment VII. There is a general age-related effect present in reaction times and error rates. The interaction of age with the target's position and the interaction of age and congruency appear to be significant in both experiments (Experiment VII and VIII). Different than in Experiment VII, a three-way interaction of age, target's position and congruency in error rates is given, indicating that the congruency effect is more pronounced for old participants for "spatially more complex" positions of the target.
In agreement with the results of Experiment V there is a significant difference

in reverberation levels in error rates, indicating higher error rates for longer reverberation times. However, no interaction of reverberation with age or any other variable can be found.
This is in line with the cited studies employing middle-age to elderly participants in shadowing tasks measuring performance in terms of accuracy [152, 101, 58, 56].

5 Conclusion

This chapter summarizes the findings of the eight conducted experiments and gives an outlook on how the examination on the intentional switching of auditory selective attention can be continued.

5.1. General Discussion and Summary

The aim of the present thesis is to examine the cognitive control mechanisms underlying auditory selective attention by considering the influence of variables that increase the complexity of an auditory scene with respect to technical aspects such as dynamic binaural hearing, room acoustics and head movements as well as those that influence the efficiency of cognitive processing. Furthermore, technical methods and tools to realize the complex auditory scenarios are evaluated with respect to the empirical findings on auditory selective attention. Step-wise the well-established dichotic-listening paradigm is extended into a "realistic" spatial listening paradigm.

In preliminary studies using the dichotic-listening paradigm, it is observed that participants can easily follow the instruction to switch auditory attention to a new auditory target. These attention switches entail switch costs in performance (longer reaction times and higher error rates). Furthermore, it is found that participants cope with the challenge of listening selectively to the relevant speaker, however, they cannot avoid processing the irrelevant information up to the level where the participant's response is influenced [45, 74].
These effects are also found when using a binaural reproduction (real sources and binaural synthesis via headphones). In a direct comparison (compare Experiment I, chapter 4.1) effects are also equally distinct. These results justify that the dichotic reproduction is successful in representing very simple spatial listening setups (source positions to the left and the right) when analyzing intentional switching of auditory selective attention.
As in dichotic results, the binaural findings clearly reveal that irrelevant informa-

tion is not necessarily unattended. When listening to a binaural reproduction of two simultaneously presented stimuli, each stimulus reaches both ears (cross-talk). The assumptions by Broadbent [20] on the filter theory, indicating an ear-based selection mechanism at a very low level of processing, are very contradictory to the findings using the binaural-listening paradigm. The late selection theory by Deutsch and Deutsch [32] is also based on an ear-wise processing, but predicts a selection taking place subsequently to perceptual processing. The present findings also disagree with this theory, since it is not conferrable to and not reasonable in real complex auditory environments. The attenuation theory which predicts a damping of processing the to be unattended information is compliant with the binaural reproduction of stimuli and is also in line with the present findings on processing the irrelevant information as far as affecting the response selection. Comparing these three theories on auditory selective attention, the attenuation theory is clearly most concordant with the present empirical findings.

On the one hand the paradigm on intentional switching of auditory selective attention is used to verify the results collected with the dichotic-listening version with these of the binaural-listening version which is closer to realistic listening. On the other hand the paradigm is also used as a measure of how plausible and authentic the binaural-reproduction techniques and adjustments regarding reverberation times are.
The extension into a binaural-listening paradigm is assessed to be successful (compare Experiment II, chapter 4.2). Auditory attention switching and the inhibition of irrelevant information in dichtoic-listening and binaural-listening lead to comparable outcomes in the performance measures of reaction time and accuracy.
The extension of the dichotic-listening paradigm offers great potential for the analysis of intentional switching of auditory attention by enabling more possibilities for spatial sources. Correlated with the degree of difficulty in localization [116, 180] performance measures vary depending on the position of target and distracting source. For a positioning on median plane participant's reaction times and accuracy drop significantly compared to other spatial positions. The arrangement of target and distractor have a great impact on the cognitive processing, especially in inhibiting the distracting sources.

Four different reproduction methods are compared using the experimental paradigm (compare Experiment II, chapter 4.2). It is common to apply localization experiments [160, 25, 180, 116, 21, 181] or direct comparisons of stimuli to evaluate the plausibility and authenticity [38, 39, 41, 42, 126, 131, 186, 119, 82, 156, 94, 19] of binaural reproduction methods. Localization and the sensitivity

to notice differences in coloration is only part of the complex functionality of our hearing system when orienting and coping in real environments. Binaural reproduction methods should be plausible in realistic environmental scenarios, as for example a "cocktail-party". Using the paradigm to intentionally switch auditory selective attention the comparison of reproduction methods is conducted in a scene of multiple talkers where the focus of attention needs to be shifted and the ability to inhibit the irrelevant information is necessary in order to fulfill the designated task.
The loss of individual information, no perfect channel separation (CTC) and possibly also the circumstances of wearing headphones diminish the ability of ignoring a distracting speaker in a spatial setup and therefore intensifies the observed effect of congruency. The performance of attention switching, however, is negligibly affected by the reproduction methods.
When using real sources performance measures are lowest level and observed effects show the closest matching to dichotic reproduction. It is assumed that the observed (partly significant) difference between the reproduction of real sources and the individual binaural reproduction results origin from the restriction of head movements for the binaural synthesis. A subsequent experiment (compare Experiment VI, chapter 4.6) gives evidence on the fact that head movements are generally very small ($< 1\,^{\circ}$) and that head movements cannot be the basis of the problematic disparity. Since the inhibitory deficit for binaural syntheses of all tested kinds is especially true for sources positioned on median plane, inaccuracy of measurements, processing or reproduction, which can never totally be eradicated, are possible reasons. On median plane, minuscule deviations in coloration can cause enormous localization errors. To assess this possibility measuring system and post processing [148], as well as latency concerns and imprecision of reproduction [65, 179] need to be evaluated. Other considerations regarding the disagreement of performance for real sources and the individual binaural synthesis are the simple fact that headphones have to be worn for the synthesis and influence the participants in some psychological way. This may be because most naive listeners are not used to listen to spatial scenes via headphones.

An anechoic reproduction fails to represent realistic listening experiences. Taking steps towards realistic environments the impact of reverberation on the intentional switching of auditory selective attention is analyzed. To be able to integrate reverberation expediently into the paradigm, the binaural-listening paradigm is extended anew, by means of longer, polysyllabic stimuli. The extension is rated favorably, however, the consequences of longer reaction times and a penalty in effect size need to be taken into account (compare chapter 4.4).
The extended binaural-listening paradigm is tested in simulated rooms with dif-

ferent reverberation times (compare chapter 4.5), showing how "reverberation has a detrimental effect on reaction times when maintaining attention to one source at a constant spatial location. However, intentionally switching the attention to a sound source at a different spatial location requires per se more attention and is more difficult that additional reverberant energy does not have any impact. Furthermore, the human ability to ignore or rather not to process the content of a distracting source is significantly influenced by reverberation"[136].

Taking into account that there is a trend towards an aging society, age-related effects should not be neglected. The hypothesis that older adults suffer from inhibitory deficits is corroborated by the studies across age groups (compare chapter 4.7 and 4.8). Furthermore, the idea of "general slowing" is confirmed in accordance with earlier findings applying the dichotic-listening paradigm [86]. Indeed, elderly listeners do not suffer losses in switching attention or inhibiting irrelevant information compared to young listeners when an additional obstruction of long reverberation time is applied. The question as to whether there is an age-related decline also with regard to auditory attention in complex environments deserves closer inspection.

5.2. Future Directions

As the title of the present thesis suggests, there is a long way from dichotic listening to listening in realistic complex environments. Considerable progress has been made in this thesis and essential insights have been gained, but there are important steps to be made to achieve the aim of the examination of the intentional switching of auditory selective attention in realistic complex environments.

Establishing a binaural-listening paradigm based on the already well-proven dichotic-listening paradigm, allowed us to examine important aspects of auditory attention in selective listening in spatial environments in the presence of multiple speakers and reverberation, while at the same time affording close experimental control. There may be other variables, such as not negligible head and body movements as well as additional distracting noise sources positioned all around the listener, that contribute to the complexity of an auditory scene, influencing the efficiency of cognitive processing. The constraint of focusing the visual cue on the monitor of the experimental setup resulted in limited headmovements. To circumvent that there is the necessity of a more liberated presentation form of the visual cue. Furthermore, the impact of visual impressions should not be underestimated, implying a multisensory paradigm. Further extensions of the binaural-listening paradigm may therefore be necessary to examine auditory selective attention in realistic complex environments. Finally, society is diverse which is why the group of participants should be widened by involving for example children and hearing-impaired listeners.

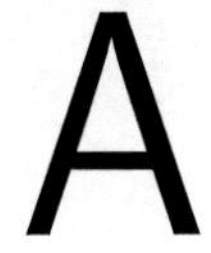

Appendix

A.1. Experiment I

Table A.1.: Experiment I: Results of repeated measures ANOVA (if needed Huynh-Feldt corrections). Main effects and Interactions in reaction times. For variable explanation, see table 4.1 and for post-hoc tests, see table A.3.

Experiment I - Reaction Times

Main effects				
Repro. Meth.		F(2,58)=1.42	p=.25	η_p^2=.05
Attent. Sw.	*	F(1,29)=43.64	p<.001	η_p^2=.60
Cong.	*	F(1,29)=6.96	p=.01	η_p^2=.19
Interactions				
Repro. Meth.*Attent. Sw.		F(2,58)=1.41	p=.25	η_p^2=.05
Repro. Meth.*Cong.		F<1		η_p^2=.03
Attent. Sw.*Cong.		F<1		η_p^2<.001
Repro. Meth.*Attent. Sw.*Cong.		F(2,58)=1.41	p=.25	η_p^2=.05

Table A.2.: Experiment I: Results of repeated measures ANOVA (if needed Huynh-Feldt corrections) Main effects and Interactions in error rates. For variable explanation, see table 4.1 and for post-hoc tests, see table A.3.

Experiment I - Error Rates

Main effects				
Repro. Meth.		F(1.16,33.69)=1.44	p=.25	η_p^2=.05
Attent. Sw.	*	F(1,29)=30.99	p<.001	η_p^2=.52
Cong.	*	F(1,29)=6.26	p=.02	η_p^2=.18
Interactions				
Repro. Meth.*Attent. Sw.		F(2,58)=1.36	p=.27	η_p^2=.05
Repro. Meth.*Cong.		F(1.11,32.17)=1.38	p=.26	η_p^2=.05
Attent. Sw.*Cong.		F<1		η_p^2=.02
Repro. Meth.*Attent. Sw.*Cong.		F<1		η_p^2=.03

Table A.3.: Experiment I: Post-hoc tests on main effects in reaction times and error rates. Asterisk without any number (*) indicates significant difference between all levels, asterisk and numbers in brackets (e.g. *(1)) indicate significant difference between present level and the level with the correspondent number. For variable explanation, see table 4.1 and for main effects and interactions see tables A.1 and A.2.

	Reproduction Method		
	Dichotic	Binaural	Real Source
Reaction Time	1015 ms	1052 ms	968 ms
Error Rates	4.0 %	4.0 %	2.8 %
	Attention Switch		
	Repetition	Switch	Switch Costs
Reaction Time	* 960 ms	* 1064 ms	104 ms
Error Rates	* 2.7 %	* 4.5 %	1.8 %
	Congruency		
	Congruent	Incongruent	Congruency Effect
Reaction Time	* 997 ms	* 1027 ms	30 ms
Error Rates	* 1.9 %	* 5.3 %	3.4 %

A.2. Experiment II

Table A.4.: Experiment II: Results of repeated measures ANOVA (if needed Huynh-Feldt corrections) Main effects and Interactions in reaction times. For variable explanation, see table 4.2 and for post-hoc tests, see table A.6.

Experiment II - Reaction Times

Main effects				
Repro. Meth.	*	F(3,92)=6.67	p<.001	η_p^2=.18
Pos.	*	F(1.66,152.29)=49.69	p<.001	η_p^2=.35
Ang.	*	F(2.89,265.92)=61.69	p<.001	η_p^2=.40
Attent. Sw.	*	F(1,92)=65.18	p<.001	η_p^2=.42
Cong.	*	F(1,92)=34.43	p<.001	η_p^2=.27
Interactions				
Pos.*Repro. Meth.	*	F(4.97,152.29)=4.01	p=.002	η_p^2=.12
Ang.*Repro. Meth.		F(8.67,265.92)=1.32	p=.23	η_p^2=.04
Attent. Sw.*Repro. Meth.		F(3,92)=1.18	p=.32	η_p^2=.04
Cong.*Repro. Meth.	*	F(3,92)=2.96	p=.04	η_p^2=.09
Pos.*Ang.	*	F(3.89,357.86)=54.93	p<.001	η_p^2=.37
Pos.*Ang.*Repro. Meth.	*	F(11.67,357.86)=2.03	p=.02	η_p^2=.06
Pos.*Attent. Sw.		F<1		η_p^2=.01
Pos.*Attent. Sw.*Repro. Meth.	*	F(5.86,179.69)=3.50	p=.003	η_p^2=.10
Ang.*Attent. Sw.	*	F(2.87,263.86)=5.21	p=.002	η_p^2=.05
Ang.*Attent. Sw.*Repro. Meth.	*	F(8.60,263.86)=2.24	p=.02	η_p^2=.07
Pos.*Ang.*Attent. Sw.		F(4.83,444.53)=1.50	p=.19	η_p^2=.02
Pos.*Ang.*Attent. Sw.*Repro. Meth.		F(14.50,444.53)=1.72	p=.05	η_p^2=.05
Pos.*Cong.	*	F(1.88,172.67)=11.01	p<.001	η_p^2=.11
Pos.*Cong.*Repro. Meth.	*	F(5.63,172.67)=2.64	p=.02	η_p^2=.08
Ang.*Cong.	*	F(2.85,262.07)=5.63	p=.001	η_p^2=.06
Ang.*Cong.*Repro. Meth.		F<1		η_p^2=.02
Pos.*Ang.*Cong.	*	F(5.18,476.12)=9.12	p<.001	η_p^2=.09
Pos.*Ang.*Cong.*Repro. Meth.	*	F(15.53,476.12)=2.04	p=.01	η_p^2=.06
Attent. Sw.*Cong.		F(1,92)=1.02	p=.32	η_p^2=.01
Attent. Sw.*Cong.*Repro. Meth.		F<1		η_p^2=.02
Pos.*Attent. Sw.*Cong.		F<1		η_p^2=.01
Pos.*Attent. Sw.*Cong.*Repro. Meth.		F(5.97,182.93)=1.66	p=.13	η_p^2=.05
Ang.*Attent. Sw.*Cong.	*	F(2.80,257.63)=8.63	p<.001	η_p^2=.09
Ang.*Attent. Sw.*Cong.*Repro. Meth.		F(8.40,257.63)=1.28	p=.25	η_p^2=.04
Pos.*Ang.*Attent. Sw.*Cong.		F(4.99,458.79)=2.29	p=.05	η_p^2=.02
Pos.*Ang.*Attent. Sw.*Cong.*Repro. Meth.		F(14.96,458.79)=1.13	p=.33	η_p^2=.04

Table A.5.: Experiment II: Results of repeated measures ANOVA (if needed Huynh-Feldt corrections) Main effects and Interactions in error rates. For variable explanation, see table 4.2 and for post-hoc tests, see table A.6.

Experiment II - Error Rates

Main effects				
Repro. Meth.	*	F(3,92)=17.93	p<.001	η_p^2=.37
Pos.	*	F(1.82,167.07)=39.42	p<.001	η_p^2=.30
Ang.	*	F(2.84,261.01)=81.69	p<.001	η_p^2=.47
Attent. Sw.		F(1,92)=2.38	p=.13	η_p^2=.03
Cong.	*	F(1,92)=430.21	p<.001	η_p^2=.82
Interactions				
Pos.*Repro. Meth.	*	F(5.45,167.07)=4.31	p=.001	η_p^2=.12
Ang.*Repro. Meth.	*	F(8.51,261.01)=3.21	p=.001	η_p^2=.10
Attent. Sw.*Repro. Meth.		F<1		η_p^2=.02
Cong.*Repro. Meth.	*	F(3,92)=19.37	p<.001	η_p^2=.39
Pos.*Ang.	*	F(4.42,406.49)=134.90	p<.001	η_p^2=.60
Pos.*Ang.*Repro. Meth.	*	F(13.26,406.49)=4.40	p<.001	η_p^2=.13
Pos.*Attent. Sw.	*	F(1.72,158.34)=6.83	p=.002	η_p^2=.07
Pos.*Attent. Sw.*Repro. Meth.		F(5.16,158.34)=2.02	p=.08	η_p^2=.06
Ang.*Attent. Sw.	*	F(2.58,237.20)=3.99	p=.01	η_p^2=.04
Ang.*Attent. Sw.*Repro. Meth.		F(7.74,237.20)=1.73	p=.10	η_p^2=.05
Pos.*Ang.*Attent. Sw.		F(4.76,438.06)=2.24	p=.053	η_p^2=.02
Pos.*Ang.*Attent. Sw.*Repro. Meth.		F(14.29,438.06)=1.55	p=.09	η_p^2=.05
Pos.*Cong.	*	F(1.68,154.67)=36.66	p<.001	η_p^2=.29
Pos.*Cong.*Repro. Meth.	*	F(5.04,154.67)=3.89	p=.002	η_p^2=.11
Ang.*Cong.	*	F(2.77,254.53)=66.25	p<.001	η_p^2=.42
Ang.*Cong.*Repro. Meth.	*	F(8.30,254.53)=2.95	p=.003	η_p^2=.09
Pos.*Ang.*Cong.	*	F(4.75,437.02)=142.81	p<.001	η_p^2=.61
Pos.*Ang.*Cong.*Repro. Meth.	*	F(14.25,437.02)=6.18	p<.001	η_p^2=.17
Attent. Sw.*Cong.	*	F(1,92)=7.87	p=.006	η_p^2=.08
Attent. Sw.*Cong.*Repro. Meth.		F<1		η_p^2<.02
Pos.*Attent. Sw.*Cong.		F(1.95,179.73)=1.76	p=.18	η_p^2=.02
Pos.*Attent. Sw.*Cong.*Repro. Meth.	*	F(5.86,179.73)=3.52	p=.003	η_p^2=.10
Ang.*Attent. Sw.*Cong.		F(2.74,252.14)=2.03	p=.12	η_p^2=.02
Ang.*Attent. Sw.*Cong.*Repro. Meth.	*	F(8.22,252.14)=2.12	p=.03	η_p^2=.07
Pos.*Ang.*Attent. Sw.*Cong.		F(4.64,426.58)=2.33	p=.05	η_p^2=.03
Pos.*Ang.*Attent. Sw.*Cong.*Repro. Meth.		F(13.91,426.58)=1.63	p=.07	η_p^2=.05

Table A.6.: Experiment II: Post-hoc tests on main effects in reaction times and error rates. Asterisk without any number (*) indicates significant difference between all levels, asterisk and numbers in brackets (e.g. *(1)) indicate significant difference between present level and the level with the correspondent number. For variable explanation, see table 4.2 and for main effects and interactions see tables A.4 and A.5.

	Reproduction Method							
	Real Source		Ind. HRTF		Non-ind. HRTF		Non-ind. CTC	
Reaction Time	*(4)	1000 ms	*(4)	1027 ms		1136 ms	*(1,2)	1236 ms
Error Rates	*	3.3 %	*(1)	7.5 %	*(1)	8.3 %	*(1)	9.0 %
	Target's Position							
	Median		Diagonal		Frontal			
Reaction Time	*	1152 ms	*	1098 ms	*	1051 ms		
Error Rates	*	9.0 %	*	6.6 %	*	5.6 %		
	Angle between target and distractor							
	45 °		90 °		135 °		180 °	
Reaction Time	*	1156 ms	*(1,4)	1075 ms	*(1)	1080 ms	*(1,2)	1091 ms
Error Rates	*(2,3)	9.3 %	*	5.6 %	*	4.5 %	*(2,3)	8.8 %
	Attention Switch							
	Repetition		Switch				Switch Costs	
Reaction Time	*	1074 ms	*	1127 ms				53 ms
Error Rates		6.8 %		7.3 %				0.5 %
	Congruency							
	Congruent		Incongruent				Congruency Effect	
Reaction Time	*	1076 ms	*	1125 ms				49 ms
Error Rates	*	2.3 %	*	11.8 %				9.5 %

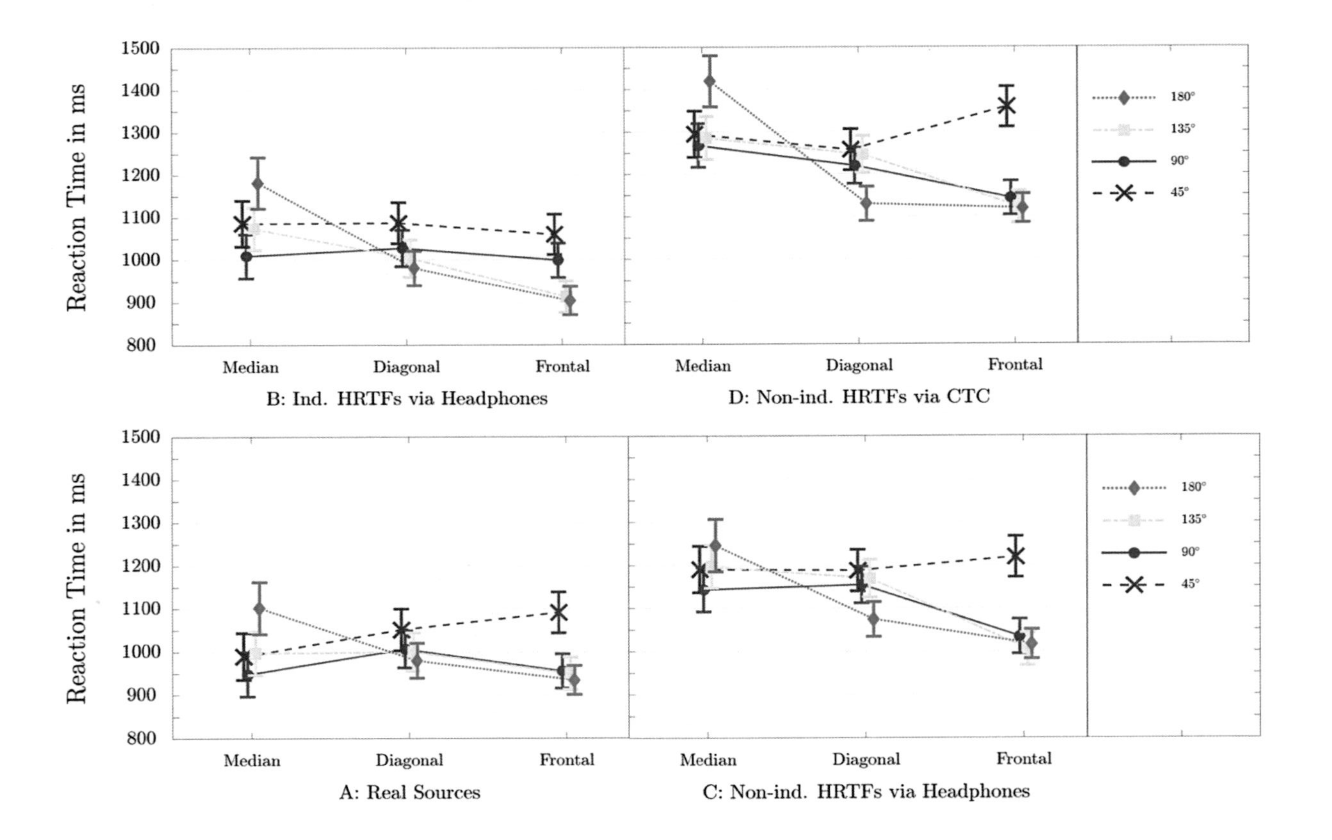

Figure A.1.: Experiment II: Reaction times (in ms) as a function of reproduction method, position and angle (RM × TPOS × ANG). Error bars indicate standard errors.

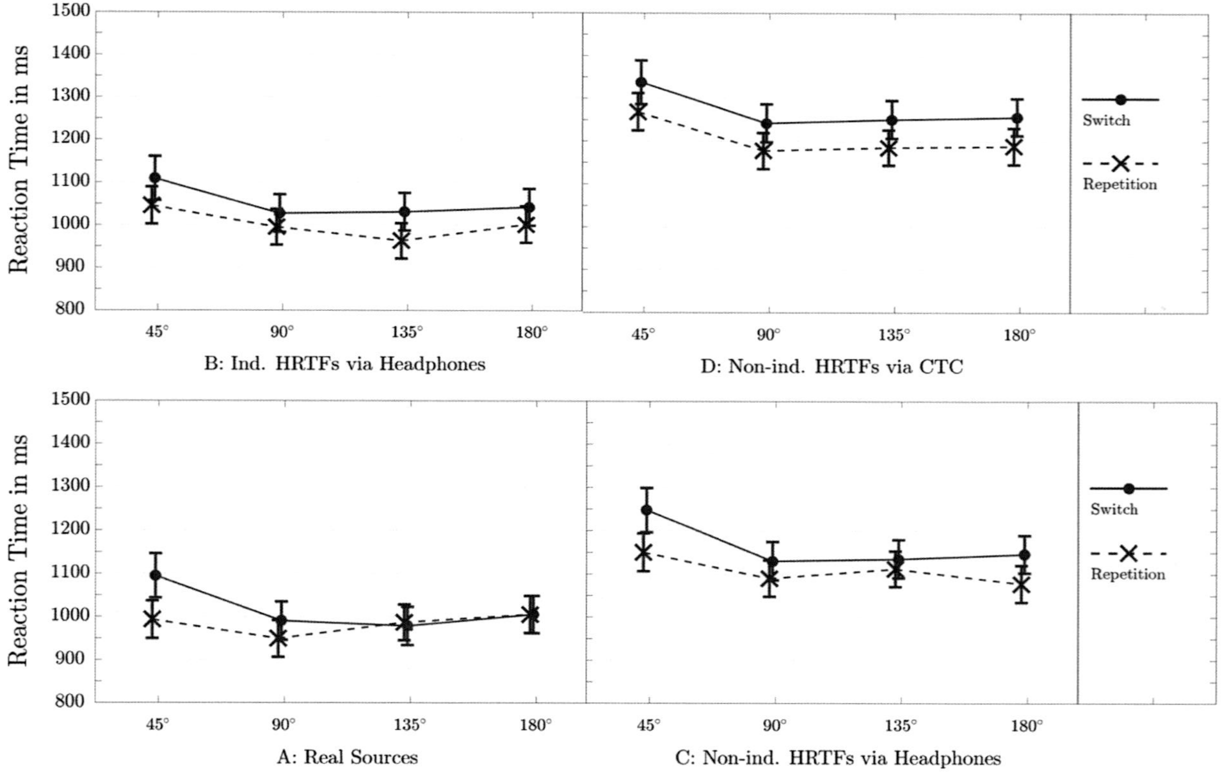

Figure A.2.: Experiment II: Reaction times (in ms) as a function of reproduction method, angle and attention switch (RM × ANG × AS). Error bars indicate standard errors.

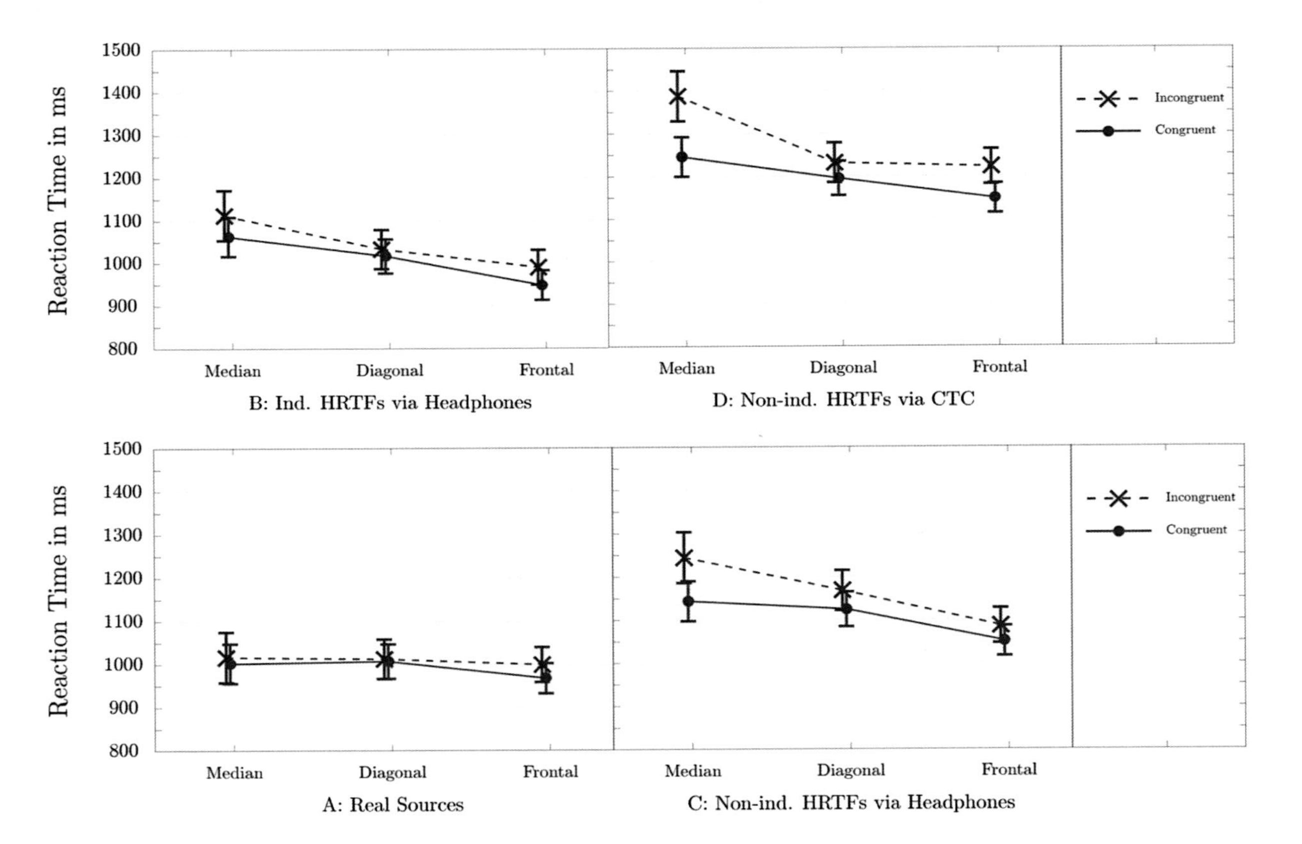

Figure A.3.: Experiment II: Reaction times (in ms) as a function of reproduction method, position and congruency (RM $\times$ TPOS $\times$ C). Error bars indicate standard errors.

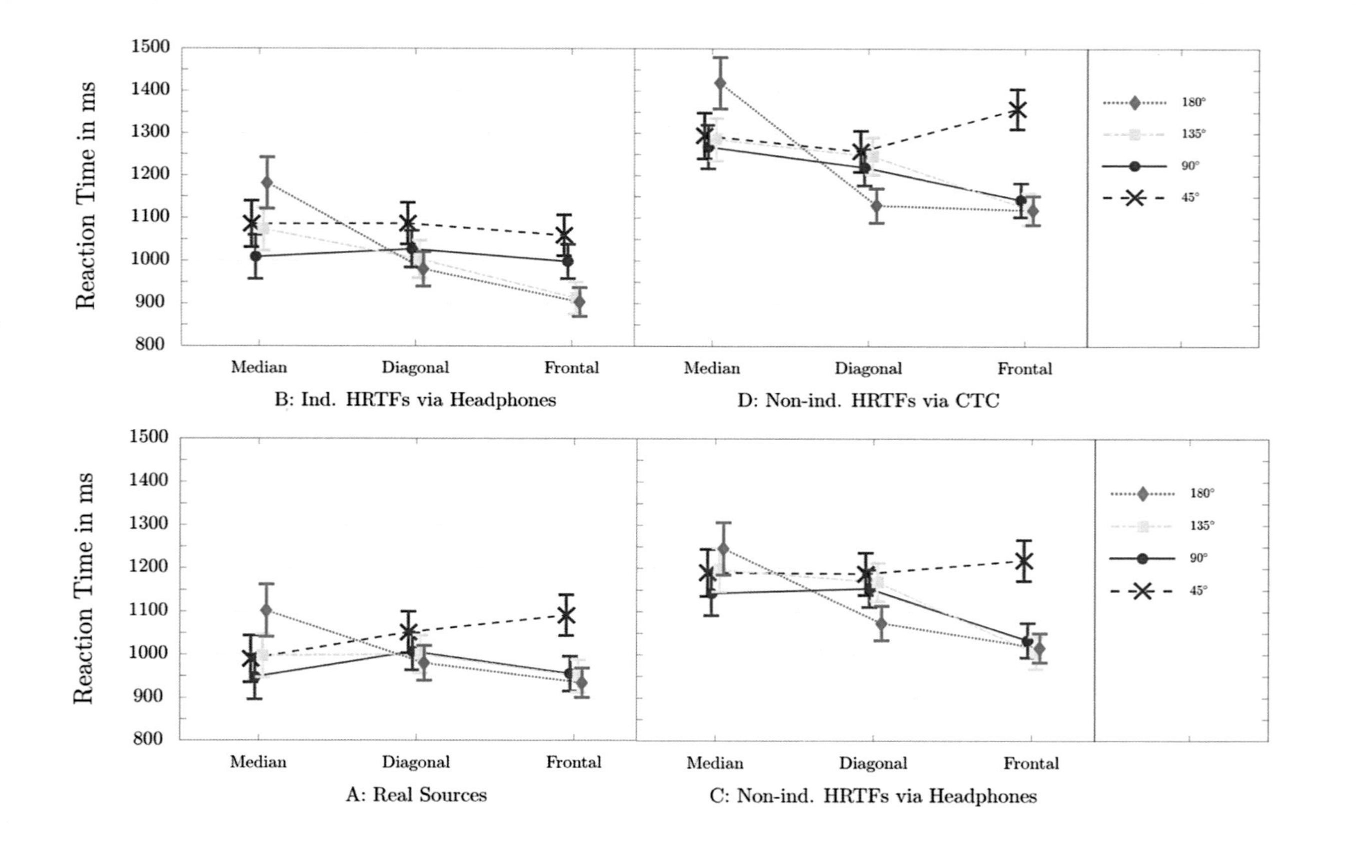

Figure A.4.: Experiment II: Error rates (in %) as a function of reproduction method, position and angle (RM × TPOS × ANG). Error bars indicate standard errors.

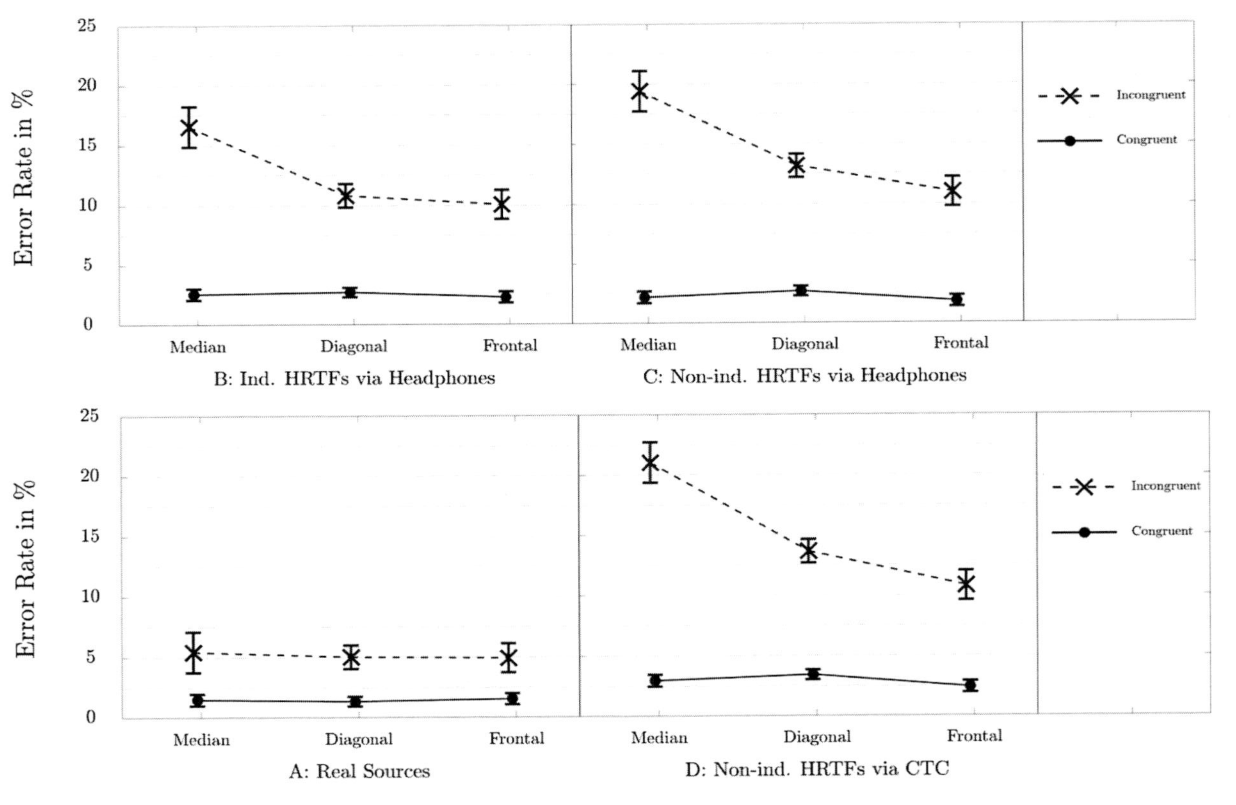

Figure A.5.: Experiment II: Error rates (in %) as a function of reproduction method, position and congruency (RM × TPOS × C). Error bars indicate standard errors.

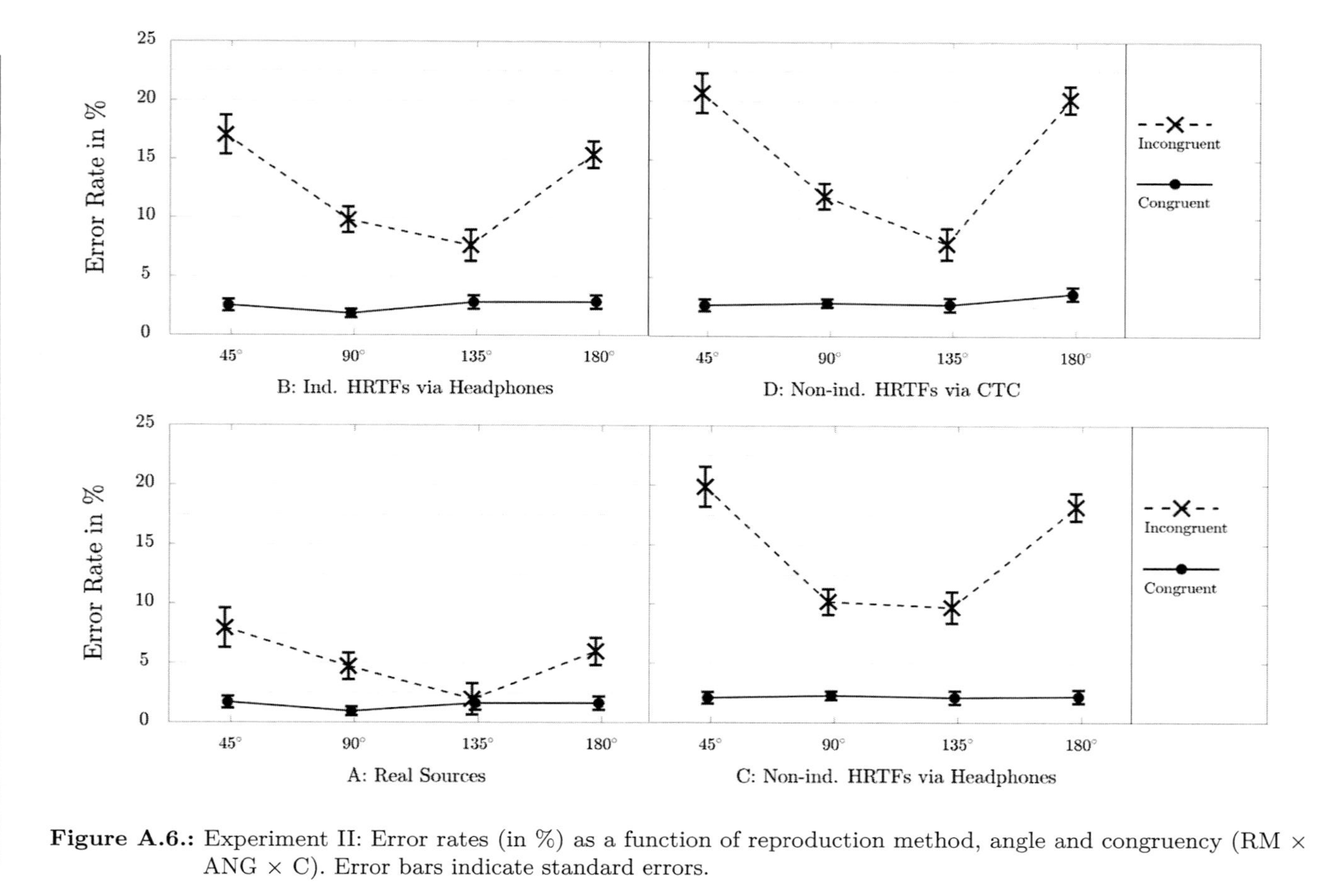

Figure A.6.: Experiment II: Error rates (in %) as a function of reproduction method, angle and congruency (RM × ANG × C). Error bars indicate standard errors.

A.3. Experiment III

Table A.7.: Experiment III: Results of repeated measures ANOVA (if needed Huynh-Feldt corrections) Main effects and Interactions in reaction times. For variable explanation, see table 4.3 and for post-hoc tests, see table A.9.

Experiment III - Reaction Times

Main effects				
Reverb.		F<1		η_p^2=.01
Pos.	*	F(1.58,70.95)=54.46	p<.001	η_p^2=.55
Attent. Sw.	*	F(1,45)=40.28	p<.001	η_p^2=.47
Cong.	*	F(1,45)=13.21	p=.001	η_p^2=.23
Interactions				
Pos.*Reverb.		F<1		η_p^2=.02
Attent. Sw.*Reverb.		F<1		η_p^2=.01
Cong.*Reverb.		F(2,45)=1.83	p=.17	η_p^2=.08
Pos.*Attent. Sw.	*	F(1.79,80.52)=4.65	p=.02	η_p^2=.09
Pos.*Attent. Sw.*Reverb.		F<1		η_p^2=.02
Pos.*Cong.	*	F(2,90)=8.28	p<.001	η_p^2=.16
Pos.*Cong.*Reverb.		F(4,90)=1.33	p=.26	η_p^2=.06
Attent. Sw.*Cong.		F<1		η_p^2=.001
Attent. Sw.*Cong.*Reverb.		F(2,45)=1.12	p=.34	η_p^2=.05
Pos.*Attent. Sw.*Cong.	*	F(2,90)=6.56	p=.002	η_p^2=.13
Pos.*Attent. Sw.*Cong.*Reverb.	*	F(4,90)=2.51	p=.047	η_p^2=.10

Table A.8.: Experiment III: Results of repeated measures ANOVA (if needed Huynh-Feldt corrections) Main effects and Interactions in error rates. For variable explanation, see table 4.3 and for post-hoc tests, see table A.9.

Experiment III - Error Rates

Main effects				
Reverb.		F(2,45)=2.68	p=.08	η_p^2=.11
Pos.	*	F(2,90)=42.14	p<.001	η_p^2=.48
Attent. Sw.		F(1,45)=2.27	p=.14	η_p^2=.05
Cong.	*	F(1,45)=271.39	p<.001	η_p^2=.86
Interactions				
Pos.*Reverb.		F<1		η_p^2=.01
Attent. Sw.*Reverb.		F<1		η_p^2=.001
Cong.*Reverb.		F(2,45)=2.46	p=.10	η_p^2=.10
Pos.*Attent. Sw.		F<1		η_p^2=.02
Pos.*Attent. Sw.*Reverb.		F<1		η_p^2=.03
Pos.*Cong.	*	F(2,90)=37.41	p<.001	η_p^2=.45
Pos.*Cong.*Reverb.		F<1		η_p^2=.02
Attent. Sw.*Cong.	*	F(1,45)=6.47	p=.02	η_p^2=.13
Attent. Sw.*Cong.*Reverb.		F(2,45)=1.33	p=.28	η_p^2=.06
Pos.*Attent. Sw.*Cong.		F<1		η_p^2=.01
Pos.*Attent. Sw.*Cong.*Reverb.		F(4,90)=2.04	p=.10	η_p^2=.08

Table A.9.: Experiment III: Post-hoc tests on main effects in reaction times and error rates. Asterisk without any number (*) indicates significant difference between all levels, asterisk and numbers in brackets (e.g. *(1)) indicate significant difference between present level and the level with the correspondent number. For variable explanation, see table 4.3 and for main effects and interactions see tables A.7 and A.8.

	Reverberation					
	Anechoic		Low Reverberation		High Reverberation	
Reaction Time		1181 ms		1141 ms		1125 ms
Error Rates		8.3 %		11.7 %		9.3 %

	Target's Position					
	Median		Diagonal		Frontal	
Reaction Time	*	1204 ms	*	1165 ms	*	1079 ms
Error Rates	*(3)	11.8 %	*(3)	10.7 %	*	6.9 %

	Attention Switch					
	Repetition		Switch		Switch Costs	
Reaction Time	*	1123 ms	*	1176 ms		54 ms
Error Rates		9.5 %		10.1 %		0.6 %

	Congruency					
	Congruent		Incongruent		Congruency Effect	
Reaction Time	*	1128 ms	*	1170 ms		42 ms
Error Rates	*	3.6 %	*	16.0 %		12.4 %

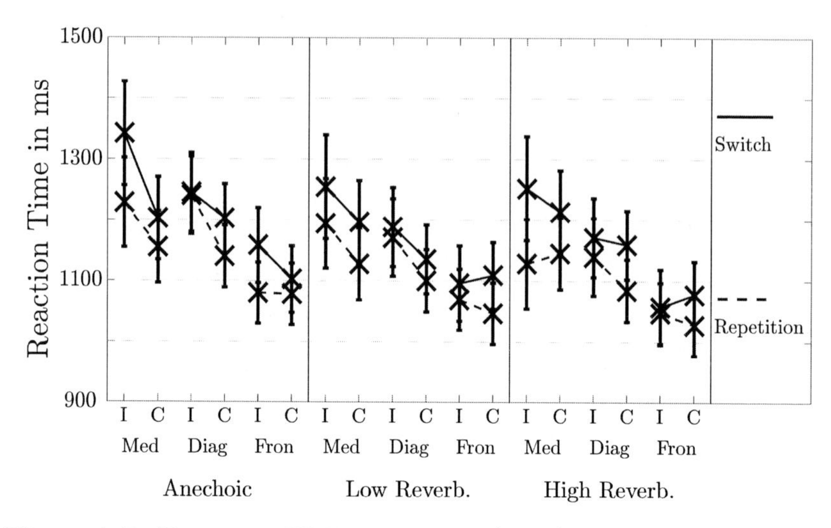

Figure A.7.: Experiment III: Reaction times (in ms) as a function of reverberation, target's position, attention switch and congruency (R × TPOS × AS × C). Error bars indicate standard errors.

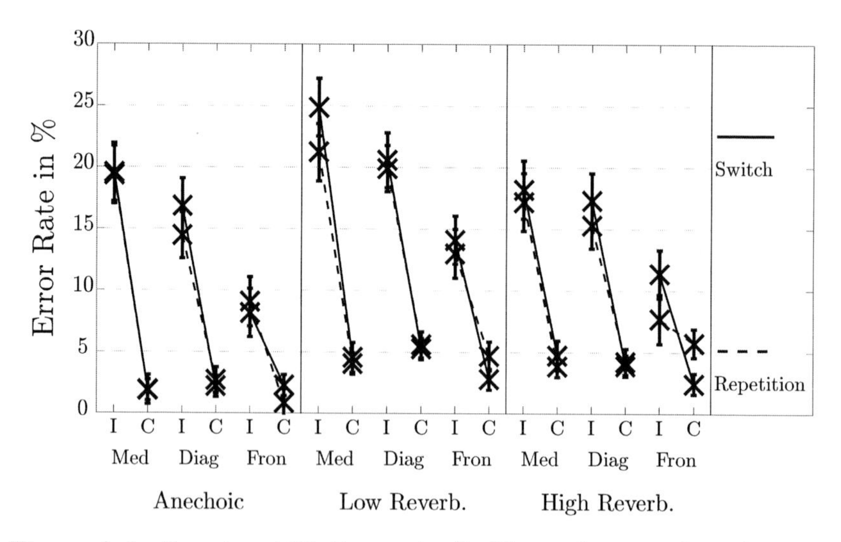

Figure A.8.: Experiment III: Error rates (in %) as a function of reverberation, target's position, attention switch and congruency (R × TPOS × AS × C). Error bars indicate standard errors.

A.4. Experiment IV

Table A.10.: Experiment IV: Results of repeated measures ANOVA (if needed Huynh-Feldt corrections) Main effects and Interactions in reaction times. For variable explanation, see table 4.4 and for post-hoc tests, see table A.12.

Experiment IV - Reaction Times

Main effects				
Parad.		F(1,46)=1.04	p=.31	η_p^2=.02
Pos.	*	F(1.40,64.36)=58.95	p<.001	η_p^2=.56
Attent. Sw.	*	F(1,46)=41.14	p<.001	η_p^2=.47
Cong.	*	F(1,46)=7.31	p=.01	η_p^2=.14
Interactions				
Pos.*Parad.		F(1.40,64.36)=1.37	p=.26	η_p^2=.03
Attent. Sw.*Parad.		F(1,46)=1.48	p=.23	η_p^2=.03
Cong.*Parad.		F(1,46)=2.01	p=.16	η_p^2=.04
Pos.*Attent. Sw.	*	F(1.78,81.77)=3.88	p=.03	η_p^2=.08
Pos.*Attent. Sw.*Parad.		F(1.78,81.77)=1.46	p=.24	η_p^2=.03
Pos.*Cong.		F<1		η_p^2=.02
Pos.*Cong.*Parad.	*	F(1.84,84.43)=4.80	p=.01	η_p^2=.10
Attent. Sw.*Cong.		F<1		η_p^2<.01
Attent. Sw.*Cong.*Parad.		F(1,46)=2.42	p=.13	η_p^2=.05
Pos.*Attent. Sw.*Cong.		F<1		η_p^2=.01
Pos.*Attent. Sw.*Cong.*Parad.		F(1.56,71.81)=2.02	p=.15	η_p^2=.04

Table A.11.: Experiment IV: Results of repeated measures ANOVA (if needed Huynh-Feldt corrections) Main effects and Interactions in error rates. For variable explanation, see table 4.4 and for post-hoc tests, see table A.12.

Experiment IV - Error Rates

Main effects				
Parad.	*	$F(1,46)=36.24$	$p<.001$	$\eta_p^2=.44$
Pos.	*	$F(2,92)=49.55$	$p<.001$	$\eta_p^2=.52$
Attent. Sw.	*	$F(1,46)=11.57$	$p=.001$	$\eta_p^2=.20$
Cong.	*	$F(1,46)=234.84$	$p<.001$	$\eta_p^2=.84$
Interactions				
Pos.*Parad.	*	$F(2,92)=4.61$	$p=.01$	$\eta_p^2=.09$
Attent. Sw.*Parad.		$F(1,46)=1.19$	$p=.28$	$\eta_p^2=.03$
Cong.*Parad.		$F<1$		$\eta_p^2=.01$
Pos.*Attent. Sw.		$F<1$		$\eta_p^2<.001$
Pos.*Attent. Sw.*Parad.		$F<1$		$\eta_p^2=.01$
Pos.*Cong.	*	$F(2,92)=26.37$	$p<.001$	$\eta_p^2=.36$
Pos.*Cong.*Parad.		$F(2,92)=1.05$	$p=.36$	$\eta_p^2=.02$
Attent. Sw.*Cong.		$F(1,46)=3.76$	$p=.06$	$\eta_p^2=.08$
Attent. Sw.*Cong.*Parad.		$F<1$		$\eta_p^2=.01$
Pos.*Attent. Sw.*Cong.		$F<1$		$\eta_p^2=.01$
Pos.*Attent. Sw.*Cong.*Parad.		$F(2,92)=2.03$	$p=.14$	$\eta_p^2=.04$

Table A.12.: Experiment IV: Post-hoc tests on main effects in reaction times and error rates. Asterisk without any number (*) indicates significant difference between all levels, asterisk and numbers in brackets (e.g. *(1)) indicate significant difference between present level and the level with the correspondent number. For variable explanation, see table 4.4 and for main effects and interactions see tables A.10 and A.11.

	Paradigm				
		New		Former	Difference
Reaction Time		1241 ms		1131 ms	110 ms
Error Rates	*	14 %	*	8 %	6 %

	Target's Position					
		Median		Diagonal		Frontal
Reaction Time	*	1251 ms	*	1204 ms	*	1104 ms
Error Rates	*(3)	13.3 %	*(3)	13.0 %	*	6.8 %

	Attention Switch				
		Repetition		Switch	Switch Costs
Reaction Time	*	1147 ms	*	1225 ms	78 ms
Error Rates	*	10.2 %	*	11.8 %	1.6 %

	Congruency				
		Congruent		Incongruent	Congruency Effect
Reaction Time	*	1169 ms	*	1203 ms	34 ms
Error Rates	*	5.3 %	*	16.7 %	11.4 %

A.5. Experiment V

Table A.13.: Experiment V: Results of repeated measures ANOVA (if needed Huynh-Feldt corrections). Main effects and Interactions in reaction times. For variable explanation, see table 4.5 and for post-hoc tests, see table A.15.

Experiment V - Reaction Times

Main effects				
Reverb.		F<1		η_p^2=.02
Pos.	*	F(1.51,33.13)=32.96	p<.001	η_p^2=.60
Attent. Sw.	*	F(1,22)=17.11	p<.001	η_p^2=.44
Cong.		F(1,22)=1.90	p=.18	η_p^2=.08
Interactions				
Reverb.*Pos.		F(3.42,75.28)=1.38	p=.25	η_p^2=.06
Reverb.*Attent. Sw.	*	F(2,44)=3.45	p=.04	η_p^2=.14
Pos.*Attent. Sw.		F(2,44)=2.96	p=.06	η_p^2=.12
Reverb.*Pos.*Attent. Sw.		F<1		η_p^2=.01
Reverb.*Cong.		F<1		η_p^2=.02
Pos.*Cong.		F(2,44)=1.00	p=.38	η_p^2=.04
Reverb.*Pos.*Cong.		F(4,88)=1.20	p=.32	η_p^2=.05
Attent. Sw.*Cong.		F<1		η_p^2=.09
Reverb.*Attent. Sw.*Cong.		F<1		η_p^2=.09
Pos.*Attent. Sw.*Cong.		F<1		η_p^2=.02
Reverb.*Pos.*Attent. Sw.*Cong.		F<1		η_p^2=.02

Table A.14.: Experiment V: Results of repeated measures ANOVA (if needed Huynh-Feldt corrections). Main effects and Interactions in error rates. For variable explanation, see table 4.5 and for post-hoc tests, see table A.15.

Experiment V - Error Rates

Main effects				
Reverb.	*	F(2,44)=3.94	p=.03	η_p^2=.15
Pos.	*	F(2,44)=23.87	p<.001	η_p^2=.52
Attent. Sw.		F<1		η_p^2=.001
Cong.	*	F(1,22)=231.76	p<.001	η_p^2=.91
Interactions				
Reverb.*Pos.		F<1		η_p^2=.01
Reverb.*Attent. Sw.		F<1		η_p^2=.01
Pos.*Attent. Sw.		F(2,44)=2.38	p=.11	η_p^2=.10
Reverb.*Pos.*Attent. Sw.		F(4,88)=2.32	p=.06	η_p^2=.10
Reverb.*Cong.	*	F(2,44)=5.38	p=.01	η_p^2=.20
Pos.*Cong.	*	F(2,44)=27.74	p<.001	η_p^2=.56
Reverb.*Pos.*Cong.		F<1		η_p^2=.02
Attent. Sw.*Cong.		F<1		η_p^2<.01
Reverb.*Attent. Sw.*Cong.		F<1		η_p^2<.01
Pos.*Attent. Sw.*Cong.		F<1		η_p^2=.04
Reverb.*Pos.*Attent. Sw.*Cong.		F(4,88)=1.58	p=.19	η_p^2=.07

Table A.15.: Experiment V: Post-hoc tests on main effects in reaction times and error rates. Asterisk without any number (*) indicates significant difference between all levels, asterisk and numbers in brackets (e.g. *(1)) indicate significant difference between present level and the level with the correspondent number. For variable explanation, see table 4.5 and for main effects and interactions see tables A.13 and A.14.

	Reverberation		
	Anechoic	Low Reverberation	High Reverberation
Reaction Time	1812 ms	1871 ms	1866 ms
Error Rates	*(3) 12.4 %	13.4 %	*(1) 14.8 %
	Target's Position		
	Median	Diagonal	Frontal
Reaction Time	* 1964 ms	* 1865 ms	* 1719 ms
Error Rates	*(3) 15.9 %	*(3) 14.9 %	* 9.8 %
	Attention Switch		
	Repetition	Switch	Switch Costs
Reaction Time	* 1799 ms	* 1900 ms	101 ms
Error Rates	13.6 %	13.5 %	-0.1 %
	Congruency		
	Congruent	Incongruent	Congruency Effect
Reaction Time	1814 ms	1885 ms	71 ms
Error Rates	* 3.7 %	* 23.4 %	19.7 %

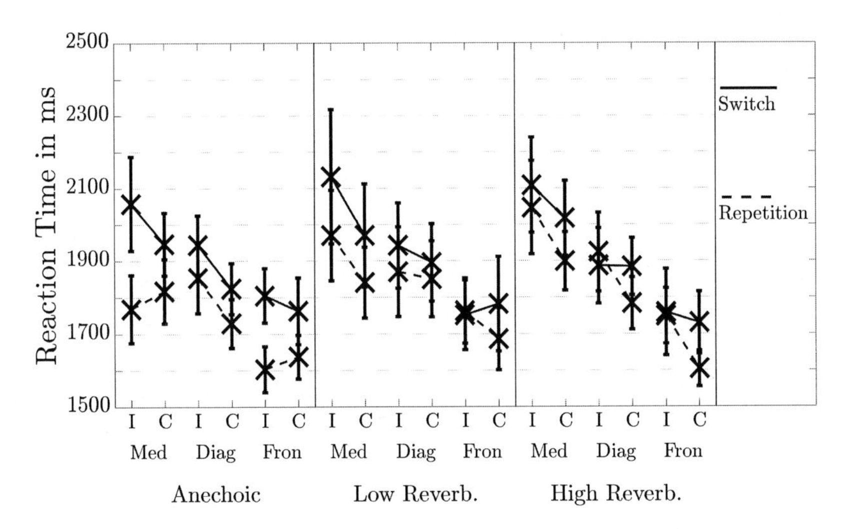

Figure A.9.: Experiment V: Reaction times (in ms) as a function of reverberation, target's position, attention switch and congruency (R × TPOS × AS × C). Error bars indicate standard errors.

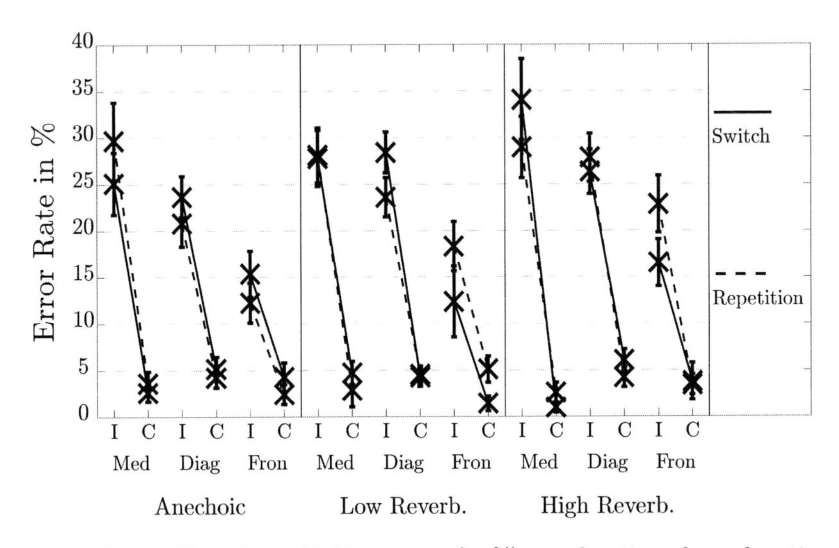

Figure A.10.: Experiment V: Error rates (in %) as a function of reverberation, target's position, attention switch and congruency (R × TPOS × AS × C). Error bars indicate standard errors.

A.6. Experiment VI

Table A.16.: Experiment VI: Results of repeated measures ANOVA (if needed Huynh-Feldt corrections). Main effects and Interactions in reaction times. For variable explanation, see table 4.6 and for post-hoc tests, see table A.18.

Experiment VI - Reaction Times

Main effects				
Rep.Meth.		F(2,44)=1.06	p=.36	η_p^2=.05
Pos.	*	F(2,44)=17.41	p<.001	η_p^2=.44
Attent. Sw.	*	F(1,22)=10.86	p=.003	η_p^2=.33
Cong.		F<1		η_p^2<.001
Interactions				
Rep.Meth.*Pos.	*	F(3.29,72.43)=3.63	p=.001	η_p^2=.14
Rep.Meth.*Attent. Sw.		F<1		η_p^2=.003
Pos.*Attent. Sw.		F(2,44)=2.82	p=.07	η_p^2=.11
Rep.Meth.*Pos.*Attent. Sw.		F(2.94,64.73)=1.31	p=.28	η_p^2=.06
Rep.Meth.*Cong.		F(1.61, 35.33)=3.64	p=.05	η_p^2=.14
Pos.*Cong.		F<1		η_p^2=.01
Rep.Meth.*Pos.*Cong.		F<1		η_p^2=.04
Attent. Sw.*Cong.		F<1		η_p^2=.02
Rep.Meth.*Attent. Sw.*Cong.		F<1		η_p^2=.02
Pos.*Attent. Sw.*Cong.		F<1		η_p^2=.01
Rep.Meth.*Pos.*Attent. Sw.*Cong.		F<1		η_p^2=.03

Table A.17.: Experiment VI: Results of repeated measures ANOVA (if needed Huynh-Feldt corrections). Main effects and Interactions in error rates. For variable explanation, see table 4.6 and for post-hoc tests, see table A.18.

Experiment VI - Error Rates

Main effects				
Rep.Meth.		F(2,44)=2.60	p=.09	η_p^2=.11
Pos.	*	F(2,44)=17.14	p<.001	η_p^2=.44
Attent. Sw.		F<1		η_p^2=.04
Cong.	*	F(1,22)=86.42	p<.001	η_p^2=.80
Interactions				
Rep.Meth.*Pos.		F(3.23,71.07)=1.19	p=.32	η_p^2=.05
Rep.Meth.*Attent. Sw.		F(2,44)=2.85	p=.07	η_p^2=.12
Pos.*Attent. Sw.		F<1		η_p^2=.001
Rep.Meth.*Pos.*Attent. Sw.		F(2.86,62.99)=1.25	p=.30	η_p^2=.05
Rep.Meth.*Cong.		F(1.67, 36.72)=2.99	p=.07	η_p^2=.12
Pos.*Cong.	*	F(2,44)=5.25	p=.009	η_p^2=.19
Rep.Meth.*Pos.*Cong.		F<1		η_p^2=.01
Attent. Sw.*Cong.		F<1		η_p^2=.04
Rep.Meth.*Attent. Sw.*Cong.		F<1		η_p^2=.01
Pos.*Attent. Sw.*Cong.		F<1		η_p^2=.04
Rep.Meth.*Pos.*Attent. Sw.*Cong.		F<1		η_p^2=.04

Table A.18.: Experiment VI: Post-hoc tests on main effects in reaction times and error rates. Asterisk without any number (*) indicates significant difference between all levels, asterisk and numbers in brackets (e.g. *(1)) indicate significant difference between present level and the level with the correspondent number. For variable explanation, see table 4.6 and for main effects and interactions see tables A.16 and A.17.

	Reproduction Method		
	Static	Quasi Static	Dynamic
Reaction Time	1912 ms	1861 ms	1892 ms
Error Rates	14.9 %	18.0 %	16.6 %
	Target's Position		
	Median	Diagonal	Frontal
Reaction Time	*(3) 1956 ms	*(3) 1911 ms	* 1798 ms
Error Rates	*(3) 20.4 %	*(3) 17.0 %	* 12.2 %
	Attention Switch		
	Repetition	Switch	Switch Costs
Reaction Time	* 1851 ms	* 1926 ms	75 ms
Error Rates	16.0 %	17.0 %	1.0 %
	Congruency		
	Congruent	Incongruent	Congruency Effect
Reaction Time	1888 ms	1889 ms	1 ms
Error Rates	* 9.5 %	* 23.5 %	14.0 %

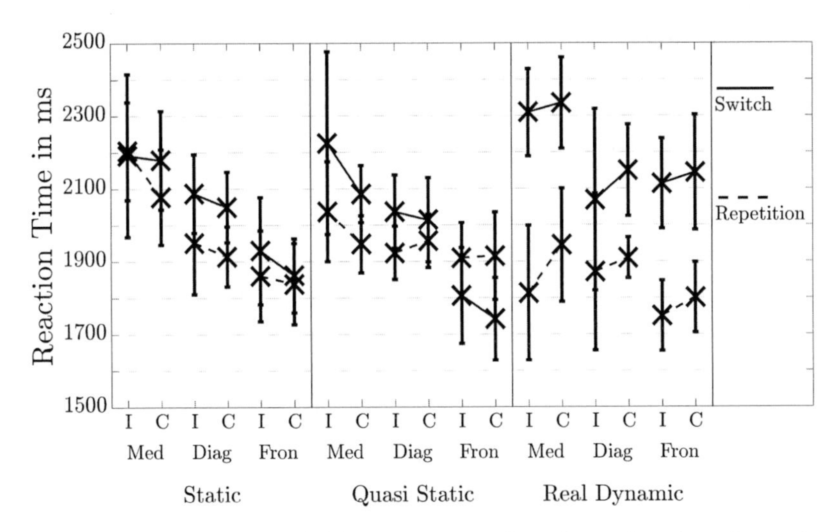

Figure A.11.: Experiment VI: Reaction times (in ms) as a function of reproduction method, target's position, attention switch and congruency (RepMeth × TPOS × AS × C). Error bars indicate standard errors.

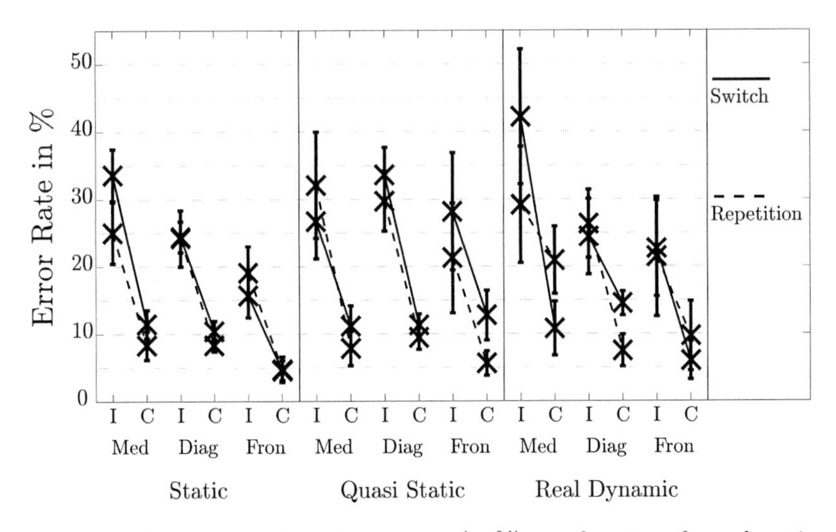

Figure A.12.: Experiment VI: Error rates (in %) as a function of reverberation, target's position, attention switch and congruency (R × TPOS × AS × C). Error bars indicate standard errors.

A.7. Experiment VII

Table A.19.: Experiment VII: Results of repeated measures ANOVA (if needed Huynh-Feldt corrections) Main effects and Interactions in reaction times. For variable explanation, see table 4.7 and for post-hoc tests, see table A.21.

Experiment VII - Reaction Times

Main effects				
Age	*	F(1,38)=23.52	$p<.001$	$\eta_p^2=.38$
Pos.	*	F(1.34,51.07)=30.93	$p<.001$	$\eta_p^2=.45$
Attent. Sw.	*	F(1,38)=21.43	$p<.001$	$\eta_p^2=.36$
Cong.	*	F(1,38)=32.44	$p<.001$	$\eta_p^2=.46$
Interactions				
Pos.*Age	*	F(1.34,51.07)=5.83	$p=.01$	$\eta_p^2=.13$
Attent. Sw.*Age		F(1,38)=2.05	$p=.16$	$\eta_p^2=.05$
Cong.*Age	*	F(1,38)=12.86	$p=.001$	$\eta_p^2=.25$
Pos.*Attent. Sw.	*	F(1.66,62.99)=4.41	$p=.02$	$\eta_p^2=.10$
Pos.*Attent. Sw.*Age		F(1.66,62.99)=1.23	$p=.30$	$\eta_p^2=.03$
Pos.*Cong.	*	F(2,76)=24.22	$p<.001$	$\eta_p^2=.39$
Pos.*Cong.*Age	*	F(2,76)=11.60	$p<.001$	$\eta_p^2=.23$
Attent. Sw.*Cong.		F<1		$\eta_p^2=.003$
Attent. Sw.*Cong.*Age		F(1,38)=1.38	$p=.25$	$\eta_p^2=.04$
Pos.*Attent. Sw.*Cong.		F<1		$\eta_p^2=.004$
Pos.*Attent. Sw.*Cong.*Age	*	F(2,76)=4.11	$p=.02$	$\eta_p^2=.10$

Table A.20.: Experiment VII: Results of repeated measures ANOVA (if needed Huynh-Feldt corrections) Main effects and Interactions in error rates. For variable explanation, see table 4.7 and for post-hoc tests, see table A.21.

Experiment VII - Error Rates

Main effects				
Age	*	F(1,38)=21.19	p<.001	η_p^2=.36
Pos.	*	F(1.76,66.87)=55.42	p<.001	η_p^2=.59
Attent. Sw.	*	F(1,38)=4.86	p=.03	η_p^2=.11
Cong.	*	F(1,38)=116.40	p<.001	η_p^2=.75
Interactions				
Pos.*Age	*	F(1.76,66.87)=7.38	p=.002	η_p^2=.16
Attent. Sw.*Age		F<1		η_p^2=.001
Cong.*Age	*	F(1,38)=7.34	p=.01	η_p^2=.16
Pos.*Attent. Sw.		F<1		η_p^2=.001
Pos.*Attent. Sw.*Age		F<1		η_p^2=.001
Pos.*Cong.	*	F(2,76)=27.19	p<.001	η_p^2=.42
Pos.*Cong.*Age		F(2,76)=2.48	p=.09	η_p^2=.06
Attent. Sw.*Cong.		F<1		η_p^2=.001
Attent. Sw.*Cong.*Age		F<1		η_p^2=.01
Pos.*Attent. Sw.*Cong.		F<1		η_p^2=.02
Pos.*Attent. Sw.*Cong.*Age		F<1		η_p^2=.002

Table A.21.: Experiment VII: Post-hoc tests on main effects in reaction times and error rates. Asterisk without any number (*) indicates significant difference between all levels, asterisk and numbers in brackets (e.g. *(1)) indicate significant difference between present level and the level with the correspondent number. For variable explanation, see table 4.7 and for main effects and interactions see tables A.19 and A.20.

	Age		
	Young	Old	Difference
Reaction Time	* 1152 ms	* 1731 ms	579 ms
Error Rates	* 8.2 %	* 16.1 %	7.9 %

	Target's Position		
	Median	Diagonal	Frontal
Reaction Time	* 1540 ms	* 1466 ms	* 1317 ms
Error Rates	* 15.8 %	* 12.8 %	* 7.9 %

	Attention Switch		
	Repetition	Switch	Switch Costs
Reaction Time	* 1396 ms	* 1486 ms	90 ms
Error Rates	* 11.7 %	* 12.6 %	0.9 %

	Congruency		
	Congruent	Incongruent	Congruency Effect
Reaction Time	* 1369 ms	* 1513 ms	144 ms
Error Rates	* 4.1 %	* 20.2 %	16.1 %

A.8. Experiment VIII

Table A.22.: Experiment VIII: Results of repeated measures ANOVA (if needed Huynh-Feldt corrections) Main effects and Interactions in reaction times. For variable explanation, see table 4.8 and for post-hoc tests, see table A.24.

Experiment VIII - Reaction Times

Main effects				
Age	*	F(1,44)=14.83	p<.001	η_p^2=.25
Reverb.		F<1		η_p^2=.01
Pos.	*	F(1.35,59.22)=21.30	p<.001	η_p^2=.33
Attent. Sw.	*	F(1,44)=23.54	p<.001	η_p^2=.35
Interactions				
Reverb.*Age		F(2,88)=1.42	p=.25	η_p^2=.03
Pos.*Age		F<1		η_p^2=.01
Attent. Sw.*Age		F<1		η_p^2=.02
Reverb.*Pos.		F<1		η_p^2=.01
Reverb.*Pos.*Age		F(3.56,156.47)=2.13	p=.09	η_p^2=.05
Reverb.*Attent. Sw.		F<1		η_p^2=.01
Reverb.*Attent. Sw.*Age		F<1		η_p^2=.02
Pos.*Attent. Sw.	*	F(2,88)=3.25	p=.04	η_p^2=.07
Pos.*Attent. Sw.*Age	*	F(2,88)=3.86	p=.03	η_p^2=.08
Reverb.*Pos.*Attent. Sw.		F<1		η_p^2=.02
Reverb.*Pos.*Attent. Sw.*Age		F<1		η_p^2=.02

Table A.23.: Experiment VIII: Results of repeated measures ANOVA (if needed Huynh-Feldt corrections) Main effects and Interactions in error rates. For variable explanation, see table 4.8 and for post-hoc tests, see table A.24.

Experiment VIII - Error Rates

Main effects				
Age	*	$F(1,44)=7.31$	$p=.01$	$\eta_p^2=.14$
Reverb.	*	$F(2,88)=6.57$	$p=.002$	$\eta_p^2=.13$
Pos.	*	$F(1.51,66.59)=52.77$	$p<.001$	$\eta_p^2=.55$
Cong.	*	$F(1,44)=368.37$	$p<.001$	$\eta_p^2=.89$
Interactions				
Reverb.*Age		$F<1$		$\eta_p^2=.01$
Pos.*Age	*	$F(1.51,66.59)=6.36$	$p=.006$	$\eta_p^2=.13$
Cong.*Age	*	$F(1,44)=25.58$	$p<.001$	$\eta_p^2=.37$
Reverb.*Pos.		$F<1$		$\eta_p^2=.01$
Reverb.*Pos.*Age		$F<1$		$\eta_p^2=.01$
Reverb.*Cong.		$F(2,88)=2.17$	$p=.12$	$\eta_p^2=.05$
Reverb.*Cong.*Age		$F<1$		$\eta_p^2=.01$
Pos.*Cong.	*	$F(1.70,74.81)=58.63$	$p<.001$	$\eta_p^2=.57$
Pos.*Cong.*Age	*	$F(1.70,74.81)=6.08$	$p=.006$	$\eta_p^2=.12$
Reverb.*Pos.*Cong.		$F<1$		$\eta_p^2=.003$
Reverb.*Pos.*Cong.*Age		$F<1$		$\eta_p^2=.004$

Table A.24.: Experiment VIII: Post-hoc tests on main effects in reaction times and error rates. Asterisk without any number (*) indicates significant difference between all levels, asterisk and numbers in brackets (e.g. *(1)) indicate significant difference between present level and the level with the correspondent number. For variable explanation, see table 4.8 and for main effects and interactions see tables A.22 and A.23.

	Age		
	Young	Old	Difference
Reaction Time	* 1836 ms	* 2255 ms	419 ms
Error Rates	* 15.0 %	* 20.9 %	5.9 %
	Reverberation		
	Anechoic	Low	High
Reaction Time	2022 ms	2041 ms	2074 ms
Error Rates	*(3) 16.9 %	*(3) 17.3 %	* 19.7 %
	Target's Position		
	Median	Diagonal	Frontal
Reaction Time	* 2132 ms	* 2057 ms	* 1947 ms
Error Rates	* 21.6 %	* 18.9 %	* 13.4 %
	Attention Switch		
	Repetition	Switch	Switch Costs
Reaction Time	* 2007 ms	* 2084 ms	77 ms
	Congruency		
	Congruent	Incongruent	Congruency Effect
Error Rates	* 4.9 %	* 31.0 %	26.1 %

List of Figures

List of Tables

Glossary

Acronyms

ANG	Spatial angle between target's and distractor's position
AS	Auditory attention switch
C	Congruency
CSI	Cue-Stimulus-Interval
CTC	Cross-Talk Cancellation
ER	Error Rate
HpTF	Headphone transfer function
HRTF	Head-related transfer function
ILD	Interaural level difference
ITD	Interaural time difference
P	Paradigm
R	Reverberation
RCI	Response-Cue-Interval
RM	Reproduction method
RT	Reaction time
TPOS	Target's spatial position

Bibliography

[1] S. Abel, C. Giguère, A. Consoli, and B. Papsin. The effect of aging on horizontal plane sound localization. *J. Acoust. Soc. Am.*, 108(2):743–752, 2000. doi: 10.1121/1.429607.

[2] M. Aceves-Fernandez, editor. *Artificial Intelligence: Emerging trends and applications.* IntechOpen, London, UK, 2018. doi: 10.5772/intechopen.71805.

[3] K. Allen, D. Alais, and S. Carlile. Speech intelligibility reduces over distance from an attended location: Evidence for an auditory spatial gradient of attention. *Percept. Psychophys.*, 71(1):164–173, 2009. doi: 10.1121/1.2407738.

[4] B. Atal and M. Schröder. Patent: 3,236,949; Apparent sound source translator, 1966.

[5] M. Bai and C. Lee. Objective and subjective analysis of effects of listening angle on crosstalk cancellation in spatial sound reproduction. *J. Acoust. Soc. Am.*, 120(4):1976–1989, 2006. doi: 10.1121/1.2257986.

[6] D. Begault, E. Wenzel, A. Lee, and M. Anderson. Direct comparison of the impact of head tracking, reverberation, and individualized head-related transfer functions on the spatial perception of a virtual speech source. In *108th Audio Eng. Soc. Conv.*, New York, NY, US, 2000.

[7] V. Best, E. Ozmeral, F. Gallun, K. Sen, and B. Shinn-Cunningham. Spatial unmasking of birdsong in human listeners: Energetic and informational factors. *J. Acoust. Soc. Am.*, 118:3766, 2005. doi: 10.1121/1.2130949.

[8] V. Best, F. Gallun, A. Ihlefeld, and B. Shinn-Cunningham. The influence of spatial separation on divided listening. *J. Acoust. Soc. Am.*, 120:1506, 2006. doi: 10.1121/1.2234849.

[9] V. Best, E. Ozmeral, and B. Shinn-Cunningham. Visually-guided attention enhances target identification in a complex auditory scene. *J. Assoc. Res. Oto.*, 8(2):294–304, 2007. doi: 10.1007/s10162-007-0073-z.

[10] V. Best, B. Shinn-Cunningham, E. Ozmeral, and N. Kopčo. Exploring the benefit of auditory spatial continuity. *J. Acoust. Soc. Am.*, 127(6): EL258–EL264, 2010. doi: 10.1121/1.3431093.

[11] J. Blauert. *Spatial Hearing - The psychophysics of human sound localization.* MIT Press, Cambridge MA, 2nd edition, 1997. ISBN 978-0262024136.

[12] J. Blauert and J. Braasch. Räumliches Hören. In S. Weinzierl, editor, *Handbuch der Audiotechnik*, pages 87–122. Springer-Verlag, BerlinHeidelberg, 2008. ISBN 978-3-540-34301-1. doi: 10.1002/bapi.200890058.

[13] J. Blauert, G. Klump, H. Hudde, J. Braasch, A. Kohlrausch, S. van de Par, H. Fastl, S. Möller, U. Jekosch, D. Hammershøi, H. Møller, I. Holube, V. Hamacher, P. Novo, J. Mourjopolos, A. Lacroix, and U. Heute. *Communication Acoustics*. Springer-Verlag, 2005. ISBN 978-3-540-27437-7.

[14] G. Boerger, J. Blauert, and P. Laws. Stereophone Kopfhörerwidergabe mit Steuerung bestimmter Übertragungsfaktoren durch Kopfdrehbewegungen. *Acust.*, 39(21-26), 1977.

[15] R. Bolia, W. Nelson, M. Ericson, and B. Simpson. A speech corpus for multitalker communications research. *J. Acoust. Soc. Am.*, 107(2):1065–1066, 2000. doi: 10.1121/1.428288.

[16] J. Bortz and R. Weber. *Statistik: für Human- und Sozialwissenschaftler*. Springer-Verlag, 6th edition, 2005. doi: 10.1007/B137571.

[17] T. Braver and D. Barch. A theory of cognitive control, aging cognition, and neuromodulation. *Neurosci. Biobehav. R.*, 26(7):809–817, 2002. ISSN 0149-7634. doi: 10.1016/S0149-7634(02)00067-2.

[18] A. S. Bregman. *Auditory Scene Analysis: The Perceptual Organization of Sound*. MIT Press, Massachusetts, 1994. doi: 10.1121/1.408434.

[19] F. Brinkmann, A. Lindau, M. Vrhovnik, and S. Weinzierl. Assessing the authenticity of individual dynamic binaural synthesis. In *Proc. of the EAA Auralization Ambisonics*, pages 62–68, 2014. ISBN 978-3-7983-2704-7.

[20] D. Broadbent. *Perception and communication*. Pergamon, Oxford, Oxford, 1958. ISBN 9781483225821.

[21] A. Bronkhorst. Localization of real and virtual sound sources. *J. Acoust. Soc. Am.*, 98(5):2542–2553, 1995. doi: 10.1121/1.413219.

[22] A. Bronkhorst. The cocktail-party problem revisited: early processing and selection of multi-talker speech. *Atten. Percept. Psycho.*, 77(5):1465–1487, 2015. doi: 10.3758/s13414-015-0882-9.

[23] D. Brungart and B. Simpson. Within-ear and across-ear interference in a dichotic cocktail party listening task: Effects of masker uncertainty. *J. Acoust. Soc. Am.*, 115(1):301–310, 2004. doi: 10.1121/1.1628683.

[24] R. Butler. Monaural and binaural localization of noise bursts vertically in the median sagittal plane. *J. Aud. Res.*, 3:230–235, 1969.

[25] R. Butler and K. Belendiuk. Spectral cues utilized in the localization of sound in the median sagittal plane. *J. Acoust. Soc. Am.*, 61:1264–1269, 1977. doi: 10.1121/1.381427.

[26] E. Cherry. Some experiments on the recognition of speech, with one and two ears. *J. Acoust. Soc. Am.*, 25(5):975–979, 1953. doi: 10.1121/1.1907229.

[27] J. F. Culling, K. Hodder, and C. Toh. Effects of reverberation on perceptual segregation of competing voices. *J. Acoust. Soc. Am.*, 114(5):2871–2876, 2003. doi: 10.1121/1.1616922.

[28] C. Darwin and R. Hukin. Effectiveness of spatial cues, prosody, and talker characteristics in selective attention. *J. Acoust. Soc. Am.*, 107(2):970–977, 2000. doi: 10.1121/1.428278.

[29] C. Darwin and R. Hukin. Effects of reverberation on spatial, prosodic, and vocal-tract size cues to selective attention. *J. Acoust. Soc. Am.*, 108(1): 335–342, 2000. doi: 10.1121/1.429468.

[30] E. David, N. Guttman, and W. van Bergeijk. Binaural interaction of high–frequency complex stimuli. *J. Acoust. Soc. Am.*, 31(6):774–782, 1959. doi: 10.1121/1.1907784.

[31] D. Deutsch. Auditory illusions, handedness, and the spatial environment. *J. Aud. Eng. Soc.*, 31(9):606–620, 1983.

[32] J. Deutsch and D. Deutsch. Attention: Some theoretical considerations. *Psychol. Rev.*, 70:80–90, 1963. doi: 10.1037/h003951.

[33] P. Dietrich, B. Masiero, and M. Vorländer. On the optimization of the multiple exponential sweep method. *J. Aud. Eng. Soc.*, 61(3):113–124, 2013.

[34] M. Dobreva, W. O'Neill, and G. Paige. Influence of aging on human sound localization. *J. Neurophysiol.*, 105(5):2471–2486, 2011. doi: 10.1152/jn.00951.2010.

[35] R. Drullman and A. Bronkhorst. Multichannel speech intelligibility and talker recognition using monaural, binaural, and three-dimensional auditory presentation. *J. Acoust. Soc. Am.*, 107(4):2224–2235, 2000. doi: 10.1121/1.428503.

[36] A. Duquesnoy. The intelligibility of sentences in quiet and in noise in aged listeners. *J. Acoust. Soc. Am.*, 74(4):1136–1144, 1983. doi: 10.1121/1.390037.

[37] J. Fels. *From children to adults: How binaural cues and ear canal impedances grow: Dissertation*, volume 5 of *Aachener Beiträge zur Akustik*. Logos Verlag Berlin Gmbh, Berlin Germany, 2008. ISBN 978-3-8325-1855-4.

[38] J. Fels, B. Masiero, J. Oberem, V. Lawo, and I. Koch. Performance of binaural technology for auditory selective attention. In *Proceedings of Acoustics 2012 Hong Kong*, volume 131, page 3317, Melville, NY, 2012. J. Acoust. Soc. Am. doi: 10.1121/1.4708412.

[39] J. Fels, M. Vorländer, B. Masiero, J. Oberem, V. Lawo, and I. Koch. Experiments on cognitive performance using binaural stimuli. In *Proc. INTER-NOISE*, 2012.

[40] J. Fels, J. Oberem, B. Karnbach, V. Lawo, and I. Koch. Comparison of dichotic and binaural reproduction in an experiment on selective auditive attention. In *Fortschritte der Akustik: AIA-DAGA 2013*, pages 1286–1289, 2013.

[41] J. Fels, J. Oberem, and B. Masiero. Experiments on authenticity and naturalness of binaural reproduction via headphones. In *21st ICA, 165th Meeting Acoust. Soc. Am., 52nd Meeting Canad. Acoust. Assoc.*, 2013. doi: 10.1121/1.4799533.

[42] J. Fels, J. Oberem, and B. Masiero. Experiments on authenticity and naturalness of binaural reproduction via headphones. *PoMA*, 19:050044, 2013. doi: 10.1121/1.4799533.

[43] J. Fels, J. Oberem, and I. Koch. Examining auditory selective attention in realistic, natural environments with an optimized paradigm. In *22nd ICA*, 2016. ISBN 978-987-24713-6-1. doi: 10.1121/2.0000321.

[44] J. Fels, J. Oberem, and I. Koch. Examining auditory selective attention in realistic, natural environments with an optimized paradigm. *PoMA*, 28(1): 050001, 2017. doi: 10.1121/2.0000321.

[45] J. Fels, J. Oberem, and I. Koch. Selective binaural attention and attention switching. In J. Blauert and J. Braasch, editors, *The Technology of Binaural Understanding*. Springer-Verlag, 2020. ISBN 978-3-030-00385-2. doi: 10.1007/978-3-030-00386-9.

[46] S. Friston and A. Steed. Measuring latency in virtual environments. *IEEE transactions on visualization and computer graphics*, 20(4):616–625, 2014.

[47] W. Gardner. *3-D audio using loudspeakers.* PhD thesis, Massachusetts Institute of Technology, Massachusetts USA, 1997.

[48] German Standard: 45631/A1. Calculation of loudness level and loudness from the sound spectrum - Zwicker method - Amendment 1: Calculation of the loudness of time-variant sound, 03.2010.

[49] S. Getzmann, C. Hanenberg, J. Lewald, M. Falkenstein, and E. Wascher. Effects of age on electrophysiological correlates of speech processing in a dynamic "cocktail-party" situation. *Front. Neuroscie.*, 9:341, 2015. ISSN 1662-453X. doi: 10.3389/fnins.2015.00341.

[50] C. Giguère and S. Abel. Sound localization: Effects of reverberation time, speaker array, stimulus frequency, and stimulus rise/decay. *J. Acoust. Soc. Am.*, 94(2):769–776, 1993. doi: 10.1121/1.408206.

[51] D. Hammershøi and H. Møller. Sound transmission to and within the human ear canal. *J. Acoust. Soc. Am.*, 100:408–427, 1996. doi: 10.1121/1.415856.

[52] W. Hartmann. Localization of sound in rooms. *J. Acoust. Soc. Am.*, 74(5): 1380–1391, 1983. doi: 10.1121/1.390163.

[53] W. Hartmann and A. Wittenberg. On the externalization of sound images. *J. Acoust. Soc. Am.*, 99:3678–3688, 1996. doi: 10.1121/1.414965.

[54] L. Hasher and R. Zacks. Working memory, comprehension, and aging: A review and a new view. volume 22 of *Psychol. Learn. Motiv.*, pages 193–225. Academic Press, 1988. doi: 10.1016/S0079-7421(08)60041-9.

[55] L. Hasher, S. Tonev, C. Lustig, and R. Zacks. Inhibitory control, environmental support, and self-initiated processing in aging. In M. Naveh-Benjamin and Moscovitch, M., Roediger, R. L., editors, *Perspectives on human memory and cognitive aging: Essaysin honour of Fergus Craik*, pages 286–297. Psychology Press, East Sussex, England, 2001. ISBN 978-0415650816.

[56] K. Helfer. Aging and the binaural advantage in reverberation and noise. *J. Speech Lang. Hear. R.*, 35(6):1394–1401, 1992. doi: 10.1044/jshr.3506.1394.

[57] K. Helfer and M. Vargo. Speech recognition and temporal processing in middle-aged women. *J. Am. Acad. Audiol.*, 20(4):264, 2009. doi: 10.3766/jaaa.20.4.6.

[58] K. Helfer and L. Wilber. Hearing loss, aging, and speech perception in reverberation and noise. *J. Speech Lang. Hear. R.*, 33(1):149–155, 1990. doi: 10.1044/jshr.3301.149.

[59] K. Helfer, C. Mason, and C. Marino. Aging and the perception of temporally-interleaved words. *Ear Hearing*, 34(2):160–167, 2013. doi: 10.1097/AUD.0b013e31826a8ea7.

[60] P. Hirsch, T. Schwarzkopp, M. Declerck, S. Reese, and I. Koch. Age-related differences in task switching and task preparation: Exploring the role of task-set competition. *Acta psychol.*, 170:66–73, 2016. doi: 10.1016/j.actpsy.2016.06.008.

[61] D. Holender. Semantic activation without conscious identification in dichotic listening, parafoveal vision, and visual masking: A survey and appraisal. *Behav. Brain Sci.*, 9(01):1–23, 1986. ISSN 1469-1825. doi: 10.1017/S0140525X00021269.

[62] K. Hugdahl. Fifty years of dichotic listening research – still going and going and.... *Brain Cognition*, 76(2):211–213, 2011. ISSN 0278-2626. doi: 10.1016/j.bandc.2011.03.006.

[63] L. Humes, J. Lee, and M. Coughlin. Auditory measures of selective and divided attention in young and older adults using single-talker competition. *J. Acoust. Soc. Am.*, 120(5):2926–2937, 2006. doi: 10.1121/1.2354070.

[64] Institute of Technical Acoustics, RWTH Aachen. ITA toolbox, 2012.

[65] Institute of Technical Acoustics, RWTH Aachen University. Virtual acoustics - a real-time auralization framework for scientific research, 2016.

[66] Y. Iwaya, Y. Suzuki, and D. Kimura. Effects of head movement on front-back error in sound localization. *Acoust. Sci. Technol.*, 24(5):322–324, 2003. doi: 10.1250/ast.24.322.

[67] K. Jost, W. de Baene, I. Koch, and M. Brass. A review of the role of cue processing in task switching. *Z. Psychol.*, 221(1):5–14, 2013. doi: 10.1027/2151-2604/a000125.

[68] G. Kidd, T. Arbogast, C. Mason, and F. Gallun. The advantage of knowing where to listen. *J. Acoust. Soc. Am.*, 118(6):3804–3815, 2005. doi: 10.1121/1.2109187.

[69] G. Kidd, C. Mason, A. Brughera, and W. Hartmann. The role of reverberation in release from masking due to spatial separation of sources for speech identification. *Acta Acust. united Ac.*, 91(3):526–536, 2005. doi: 10.1121/1.4809166.

[70] A. Kiesel and I. Koch. Wahrnehmung und aufmerksamkeit. In A. Kiesel and H. Spada, editors, *Lehrbuch Allgemeine Psychologie*, pages 35–120. Hogrefe & Huber Publishers, 2018. ISBN 9783456856063.

[71] A. Kiesel, M. Steinhauser, M. Wendt, M. Falkenstein, K. Jost, A. Philipp, and I. Koch. Control and interference in task switching—a review. *Psychol. Bull.*, 136(5):849–874, 2010. ISSN 1939-1455. doi: 10.1037/a0019842.

[72] I. Koch and V. Lawo. Exploring temporal dissipation of attention settings in auditory task switching. *Atten. Percept. Psycho.*, 76:73–80, 2014. doi: 10.3758/s13414-013-0571-5.

[73] I. Koch and V. Lawo. The flip side of the auditory spatial selection benefit. *Exp. Psychol.*, 62(1):66–74, 2015. doi: 10.1027/1618-3169/a000274.

[74] I. Koch, V. Lawo, J. Fels, and M. Vorländer. Switching in the cocktail party: Exploring intentional control of auditory selective attention. *J. Exp. Psychol. Human*, 37(4):1140–1147, 2011. doi: 10.1037/a0022189.

[75] A. Kramer and D. Madden. Attention. In F. I. M. Craik and T. A. Salthouse, editors, *The handbook of aging and cognition*, pages 189–249. Psychology Press, New York, NY, 2008. ISBN 978-0805859904.

[76] A. Kramer, S. Hahn, and D. Gopher. Task coordination and aging: explorations of executive control processes in the task switching paradigm. *Acta psychol.*, 101(2–3):339–378, 1999. doi: 10.1016/S0001-6918(99)00011-6.

[77] J. Kray. Task-set switching under cue-based versus memory-based switching conditions in younger and older adults. *Brain Res.*, 1105(1):83–92, 2006. ISSN 0006-8993. doi: 10.1016/j.brainres.2005.11.016.

[78] J. Kray and U. Lindenberger. Adult age differences in task switching. *Psychol. Aging*, 15(1):126–147, 2000. doi: 10.1037/0882-7974.15.1.126.

[79] J. Kray, J. Eber, and J. Karbach. Verbal self-instructions in task switching: a compensatory tool for action-control deficits in childhood and old age? *Developmental Sci.*, 11(2):223–236, 2008. ISSN 1467-7687. doi: 10.1111/j.1467-7687.2008.00673.x.

[80] J. Kray, J. Karbach, and A. Blaye. The influence of stimulus-set size on developmental changes in cognitive control and conflict adaptation. *Acta psychol.*, 140(2):119–128, 2012. doi: 10.1016/j.actpsy.2012.03.005.

[81] J. Lachter, K. Forster, and E. Ruthruff. Forty-five years after broadbent (1958): still no identification without attention. *Psychol. Rev.*, 111(4):880, 2004. doi: 10.1037/0033-295X.111.4.880.

[82] E. Langendijk and A. Bronkhorst. Fidelity of three-dimensional-sound reproduction using a virtual auditory display. *J. Acoust. Soc. Am.*, 107(1): 528–537, 2000. doi: 10.1121/1.428321.

[83] M. Lavandier and J. Culling. Speech segregation in rooms: Effects of reverberation on both target and interferer. *J. Acoust. Soc. Am.*, 122(3): 1713–1723, 2007. doi: 10.1121/1.2764469.

[84] M. Lavandier and J. Culling. Speech segregation in rooms: Monaural, binaural, and interacting effects of reverberation on target and interferer. *J. Acoust. Soc. Am.*, 123(4):2237–2248, 2008. doi: 10.1121/1.2871943.

[85] V. Lawo. *Exploring intentional control in auditory attention switching.* Dissertation, RWTH Aachen University, Aachen Germany, 2014.

[86] V. Lawo and I. Koch. Dissociable effects of auditory attention switching and stimulus–response compatibility. *Psychol. Res.*, 78(3):379–386, 2014. doi: 10.1007/s00426-014-0545-9.

[87] V. Lawo and I. Koch. Examining age-related differences in auditory attention control using a task-switching procedure. *J. Gerontol. B. - Psychol.*, 69:237–244, 2014. doi: 10.1093/geronb/gbs107.

[88] V. Lawo and I. Koch. Attention and action: The role of response mappings in auditory attention switching. *J. Cogn. Psychol.*, 27(2):194–206, 2015. doi: 10.1080/20445911.2014.995669.

[89] V. Lawo, J. Fels, J. Oberem, and I. Koch. Intentional attention switching in dichotic listening: Exploring the efficiency of nonspatial and spatial selection. *Q. J. Exp. Psychol.*, 67(10):2010–2024, 2014. doi: 10.1080/17470218.2014.898079.

[90] T. Lentz. Dynamic crosstalk cancellation for binaural synthesis in virtual reality environments. *J. Audio Eng. Soc.*, 54(4):283–294, 2006.

[91] T. Lentz. *Binaural technology for virtual reality: Dissertation*, volume 6 of *Aachener Beiträge zur Akustik*. Logos Verlag Berlin Gmbh, Berlin Germany, 2007. ISBN 978-3-8325-1935-3.

[92] T. Lentz, I. Assenmacher, and J. Sokoll. Performance of spatial audio using dynamic cross-talk cancellation. In *119th Audio Eng. Soc. Conv.*, New York, NY, US, 2005.

[93] L. Li, M. Daneman, J. Qi, and B. Schneider. Does the information content of an irrelevant source differentially affect spoken word recognition in younger and older adults? *J. Exp. Psychol. Human*, 30(6):1077–1091, 2004. doi: 10.1037/0096-1523.30.6.1077.

[94] A. Lindau, T. Hohn, and S. Weinzierl. Binaural resynthesis for comparative studies of acoustical environments. In *122nd Audio Eng. Soc. Conv.*, New York, NY, US, 2007.

[95] Lord Rayleigh. On our perception of sound direction. *Philos. Mag.*, 13: 214–232, 1907.

[96] C. Lustig, L. Hasher, and R. Zacks. Inhibitory deficit theory: Recent developmentsin a "new view". In D. S. Gorfein and C. M. MacLeod, editors, *Inhibition in cognition*, pages 145–162. American Psychological Association, Washington, D.C., USA, 2007. ISBN 978-1-59147-930-7.

[97] P. Mackensen. *Auditive localization. Head movements, an additional cue in localization.* Dissertation, Technische Universität Berlin, Berlin Germany, 2004.

[98] P. Majdak, P. Balazs, and B. Laback. Multiple exponential sweep method for fast measurement of head-related transfer functions. *J. Audio Eng. Soc.*, 55:623–637, 2007.

[99] P. Majdak, B. Masiero, and J. Fels. Sound localization in individualized and non-individualized crosstalk cancellation systems. *J. Acoust. Soc. Am.*, 133(4):2055–2068, 2013. doi: 10.1121/1.4792355.

[100] N. Marrone, C. Mason, and G. Kidd. Evaluating the benefit of hearing aids in solving the cocktail party problem. *Trends Amplif.*, 12(4):300–315, 2008. doi: 10.1177/1084713808325880.

[101] N. Marrone, C. Mason, and G. Kidd. Tuning in the spatial dimension: Evidence from a masked speech identification task. *J. Acoust. Soc. Am.*, 124(2):1146–1158, 2008. doi: 10.1121/1.2945710.

[102] B. Masiero. *Individualized binaural technology. Measurement, Equalization and Subjective Evaluation: Dissertation*, volume 13 of *Aachener Beiträge zur Akustik*. Logos Verlag Berlin Gmbh, Berlin Germany, 2012. ISBN 978-3-8325-3274-1.

[103] B. Masiero and J. Fels. Perceptually robust headphone equalization for binaural reproduction. In *130th Audio Eng. Soc. Conv.*, page 8388, New York, NY, US, 2011.

[104] U. Mayr. Age differences in the selection of mental sets: the role of inhibition, stimulus ambiguity, and response-set overlap. *Psychol. Aging*, 16(1):96–109, 2001. doi: 10.1037/0882-7974.16.1.96.

[105] J. McDowd and R. Shaw. Attention and aging: A functional perspective. In F. Craik and T. Salthouse, editors, *The handbook of aging and cognition*, volume IX, page 755. Lawrence Erlbaum Associates Publishers, Mahwah, NJ, US, 2000. ISBN 978-0805859904.

[106] S. Mehrgardt and V. Mellert. Transformation characteristics of the external human ear. *J. Acoust. Soc. Am.*, 61(6):1567–1576, 1977. doi: 10.1121/1.381470.

[107] N. Meiran. Reconfiguration of processing mode prior to task performance. *J. Exp. Psychol. Learn.*, 22(6):1423–1442, 1996. doi: 10.1037/0278-7393.22.6.1423.

[108] N. Meiran, A. Gotler, and A. Perlman. Old age is associated with a pattern of relatively intact and relatively impaired task-set switching abilities. *The J. Gerontol. B. - Psychol.*, 56(2):88–102, 2001. doi: 10.1093/geronb/56.2.P88.

[109] J. Middlebrooks. Individual differences in external-ear transfer functions reduced by scaling in frequency. *J. Acoust. Soc. Am.*, 106:1480–1492, 1999. doi: 10.1121/1.427176.

[110] J. Middlebrooks, J. Makous, and D. Green. Directional sensitivity of sound-pressure levels in the human ear canal. *J. Acoust. Soc. Am.*, 86(1):89–108, 1989. doi: 10.1121/1.398224.

[111] P. Minnaar, S. Olesen, F. Christensen, and H. Møller. The importance of head movements for binaural room synthesis. In *Proc. ICAD*, 2001.

[112] H. Møller. Reproduction of artificial-head recordings through loudspeakers. *J. Aud. Eng. Soc.*, 37(1/2):30–33, 1989.

[113] H. Møller, D. Hammershøi, J. Hundebøll, and C. Jensen. Transfer characteristics of headphones: measurements on 40 human subjects. In *92nd Audio Eng. Soc. Conv.*, 1992.

[114] H. Møller, D. Hammershøi, C. Jensen, and M. Sørensen. Transfer characteristics of headphones measured on human ears. *J. Aud. Eng. Soc.*, 43: 203–217, 1995.

[115] H. Møller, C. Jensen, D. Hammershøi, and M. Sørensen. Using a typical human subject for binaural recording. In *100th Audio Eng. Soc. Conv.*, 1996.

[116] H. Møller, M. F. Sørensen, C. Jensen, and D. Hammershøi. Binaural technique: Do we need individual recordings? *J. Audio Eng. Soc.*, 44: 451–469, 1996.

[117] T. Mondor, R. Zatorre, and N. Terrio. Constraints on the selection of auditory information. *J. Exp. Psychol. Human*, 24(1):66, 1998. doi: 10.1037/0096-1523.24.1.66.

[118] S. Monsell. Task switching. *Trends Cogn. Sci.*, 7(3):134–140, 2003. doi: 10.1016/S1364-6613(03)00028-7.

[119] A. Moore, A. Tew, and R. Nicol. An initial validation of individualized crosstalk cancellation filters for binaural perceptual experiments. *J. Audio Eng. Soc.*, 58(1/2):36–45, 2010.

[120] B. Moore. *An introduction to the psychology of hearing.* Academic Press, San Diego and USA, 5th edition, 2003. ISBN 978-90-04-25242-4.

[121] N. Moray. Attention in dichotic listening: Affective cues and the influence of instructions. *Q. J. Exp. Psychol.*, 11(1):56–60, 1959. doi: 10.1080/17470215908416289.

[122] A. Nábělek and P. Robinson. Monaural and binaural speech perception in reverberation for listeners of various ages. *J. Acoust. Soc. Am.*, 71(5): 1242–1248, 1982. doi: 10.1121/1.387773.

[123] NaturalPoint Inc. Optitrack, 2016.

[124] J. Oberem. Documentation of supplementary data-set on listening experiments published in the PhD-thesis: "Examining auditory selective attention: From dichotic towards realistic environments". *RWTH Aachen University*, 2020. doi: 10.18154/RWTH-2020-02899.

[125] J. Oberem and J. Fels. Speech material for a paradigm on the intentional switching of auditory selective attention. *RWTH Aachen University*, 2020. doi: 10.18154/RWTH-2020-02105.

[126] J. Oberem and B. Karnbach. Comparison of dichotic and binaural reproduction in an experiment on auditory selective attention. In *Poster Student Conference Prague*, 2013.

[127] J. Oberem, B. Masiero, and J. Fels. Authenticity and naturalness of binaural reproduction via headphones regarding different equalization methods. In *Fortschritte der Akustik: AIA-DAGA 2013*, pages 1624–1626, 2013.

[128] J. Oberem, V. Lawo, I. Koch, and J. Fels. Evaluation of experiments on auditory selective attention in an anechoic environment and a reverberant room with nonindividual binaural reproduction. In *Fortschritte der Akustik: DAGA - DGA 2014*, pages 582–583, 2014.

[129] J. Oberem, V. Lawo, I. Koch, and J. Fels. Intentional switching in auditory selective attention: Exploring different binaural reproduction methods in an anechoic chamber. *Acta Acust. united Ac.*, 100(6):1139–1148, 2014. doi: 10.3813/AAA.918793.

[130] J. Oberem, S. Wang, and J. Fels. Exploring age effects in auditory selective attention with a binaural reproduction method. In *Fortschritte der Akustik: DAGA 2015*, pages 1460–1461, 2015.

[131] J. Oberem, B. Masiero, and J. Fels. Experiments on authenticity and plausibility of binaural reproduction via headphones employing different recording methods. *Appl. Acoust.*, 114:71–78, 2016. ISSN 0003-682X. doi: 10.1016/j.apacoust.2016.07.009.

[132] J. Oberem, I. Koch, and J. Fels. Examining auditory selective attention in reverberant environments. In *Acoustics '17 Boston: 173rd Meeting Acoust. Soc. Am., 8th Forum Acusticum*, 2017.

[133] J. Oberem, I. Koch, and J. Fels. Intentional switching in auditory selective attention: Exploring age-related effects in a spatial setup requiring speech perception. *Acta Psychol.*, 177:36–43, 2017. doi: 10.1016/j.actpsy.2017.04.008.

[134] J. Oberem, J. Seibold, I. Koch, and J. Fels. Exploring influences on auditory selective attention by a static and a dynamic binaural reproduction. In *Fortschritte der Akustik: DAGA 2017*, pages 1154–1155, 2017.

[135] J. Oberem, J.-G. Richter, D. Setzer, J. Seibold, I. Koch, and J. Fels. Experiments on localization accuracy with non-individual and individual HRTFs comparing static and dynamic reproduction methods. In *Fortschritte der Akustik: DAGA 2018*, pages 702–705, 2018.

[136] J. Oberem, J. Seibold, I. Koch, and J. Fels. Intentional switching in auditory selective attention: Exploring attention shifts with different reverberation times. *Hearing Res.*, 359:32–39, 2018. doi: 10.1016/j.heares.2017.12.013.

[137] J. Oberem, J. Seibold, I. Koch, and J. Fels. Examining age effects in auditory selective attention in reverberant environments. In *Fortschritte der Akustik: DAGA 2019*, 2019.

[138] W. H. Organization. Report of the informal working group on prevention of deafness and hearing impairment programme planning, Geneva, 18-21 june 1991. 1991.

[139] R. Otte, M. Agterberg, M. Wanrooij, A. Snik, and A. van Opstal. Age-related hearing loss and ear morphology affect vertical but not horizontal sound-localization performance. *JARO*, 14(2):261–273, 2013. doi: 10.1007/s10162-012-0367-7.

[140] H. Pashler. *The psychology of attention.* MIT Press, 1998. ISBN 9780262161657.

[141] F. Pausch, L. Aspöck, M. Vorländer, and J. Fels. An extended binaural real-time auralization system with an interface to research hearing aids for experiments on subjects with hearing loss. *Trends Hear.*, 22:1–32, 2018. doi: 10.1177/2331216518800871.

[142] J. Pedersen and P. Minnaar. Evaluation of a 3D-audio system with head tracking. In *120th Audio Eng. Soc. Conv.*, New York, NY, US, 2006.

[143] S. Perrett and W. Noble. The contribution of head motion cues to localization of low-pass noise. *Atten. Percept. Psycho.*, 59:1018–1026, 1997. doi: 10.3758/BF03205517.

[144] M. Pichora-Fuller. Processing speed and timing in aging adults: psychoacoustics, speech perception, and comprehension. *Int. J. Audiol.*, 42(sup1): 59–67, 2003. doi: 10.3109/14992020309074625.

[145] M. Pichora-Fuller and G. Singh. Effects of age on auditory and cognitive processing: Implications for hearing aid fitting and audiologic rehabilitation. *Trends Amplif.*, 10(1):29–59, 2006. doi: 10.1177/108471380601000103.

[146] B. Rakerd and W. Hartmann. Localization of sound in rooms, II: The effects of a single reflecting surface. *J. Acoust. Soc. Am.*, 78(2):524–533, 1985. doi: 10.1121/1.392474.

[147] B. Rakerd and W. Hartmann. Localization of sound in rooms, III: Onset and duration effects. *J. Acoust. Soc. Am.*, 80:1695–1706, 1986. doi: 10.1121/1.394282.

[148] J.-G. Richter. *Fast Measurement of individual Head-related transfer functions: Dissertation*, volume 30 of *Aachener Beiträge zur Akustik*. LOGOS VERLAG BERLIN, Berlin Germany, 2019. ISBN 978-3-8325-4906-0.

[149] J.-G. Richter, G. Behler, and J. Fels. Evaluation of a fast hrtf measurement system. In *140th Audio Eng. Soc. Conv.*, page 9498, New York, NY, US, 2016.

[150] M. Rivenez, A. Guillaume, L. Bourgeon, and C. Darwin. Effect of voice characteristics on the attended and unattended processing of two concurrent messages. *Eur. J. Cogn. Psychol.*, 20(6):967–993, 2008. doi: 10.1080/09541440701686201.

[151] W. Rogers. Attention and aging. In D. Park and N. Schwarz, editors, *Cognitive aging: A primer.*, volume XIII, page 292. Psychology Press, New York, NY, US, 2000. ISBN 978-0863776922.

[152] D. Ruggles and B. Shinn-Cunningham. Spatial selective auditory attention in the presence of reverberant energy: Individual differences in normal-hearing listeners. *JARO*, 12:395–405, 2011. doi: 10.1007/s10162-010-0254-z.

[153] T. Salthouse. The processing-speed theory of adult age differences in cognition. *Psychol. Rev.*, 103:403–428, 1996. doi: 10.1037/0033-295X.103.3.403.

[154] T. Salthouse. Aging and measures of processing speed. *Biol. Psychol.*, 54 (1):35–54, 2000. ISSN 0301-0511. doi: 10.1016/S0301-0511(00)00052-1.

[155] T. Salthouse, N. Fristoe, K. McGuthry, and D. Hambrick. Relation of task switching to speed, age, and fluid intelligence. *Psychol. Aging*, 13(3): 445–461, 1998. doi: 10.1037/0882-7974.13.3.445.

[156] Z. Schärer and A. Lindau. Evaluation of equalization methods for binaural signals. In *126th Audio Eng. Soc. Conv.*, New York, NY, US, 2009.

[157] A. Schmitz. *Naturgetreue Wiedergabe kopfbezogener Schallaufnahmen über zwei Lautsprecher mit Hilfe eines Übersprechkompensators*. Dissertation, RWTH Aachen University, Aachen Germany, 1993.

[158] A. Schmitz. Ein neues digitales Kunstkopfmesssystem. *Acta Acust. united Ac.*, 81(4):416–420, 1995.

[159] D. Schröder. *Physically Based Real-Time Auralization of Interactive Virtual Environments: Dissertation*, volume 11 of *Aachener Beiträge zur Akustik*. Logos Verlag Berlin Gmbh, Berlin Germany, 2012. ISBN 978-3-8325-2458-6.

[160] C. Searle, L. Braida, D. Cuddy, and M. Davis. Binaural pinna disparity: another auditory localization cue. *J. Acoust. Soc. Am.*, 57:448–455, 1975. doi: 10.1121/1.380442.

[161] J. Seibold. *Examining independently switching components of auditory task sets: Towards a general mechanism of multicomponent switching*. Dissertation, RWTH Aachen University, Aachen Germany, 2018.

[162] J. Seibold, S. Nolden, J. Oberem, J. Fels, and I. Koch. Intentional preparation of auditory attention-switches: Explicit cueing and sequential switch-predictability. *Q. J. Exp. Psychol.*, 71(6):1382–1395, 2018. doi: 10.1080/17470218.2017.1344867.

[163] J. Seibold, S. Nolden, J. Oberem, J. Fels, and I. Koch. Auditory attention switching and judgment switching: Exploring multicomponent task representations. *Atten. Percept. Psycho.*, 80(7):1823–1832, 2018. doi: 10.3758/s13414-018-1557-0.

[164] J. Seibold, S. Nolden, J. Oberem, J. Fels, and I. Koch. The binding of an auditory target location to a judgement: A two-component switching approach. *Q. J. Exp. Psychol.*, 2019. doi: 10.1177/1747021819829422.

[165] B. Shinn-Cunningham. Object-based auditory and visual attention. *Trends Cogn. Sci.*, 12(5):182–186, 2008. doi: 10.1016/j.tics.2008.02.003.

[166] G. Singh, M. Pichora-Fuller, and B. Schneider. Time course and cost of misdirecting auditory spatial attention in younger and older adults. *Ear Hearing*, 34(6):711–721, 2013. doi: 10.1097/AUD.0b013e31829bf6ec.

[167] T. Takeuchi, P. Nelson, and H. Hamada. Robustness to head misalignment of virtual sound imaging systems. *J. Acoust. Soc. Am.*, 109(3):958–971, 2001. doi: 10.1121/1.1349539.

[168] W. Thurlow and P. Runge. Effect of induced head movements on localization of direction of sounds. *J. Acoust. Soc. Am.*, 42(2):480–488, 1967. doi: 10.1121/1.1910604.

[169] A. Treisman. Contextual cues in selective listening. *Q. J. Exp. Psychol.*, 12(4):242–248, 1960. doi: 10.1080/17470216008416732.

[170] P. Tun and M. Lachman. Age differences in reaction time and attention in a national telephone sample of adults: education, sex, and task complexity matter. *Dev. Psychol.*, 44(5):1421, 2008. doi: 10.1037/a0012845.

[171] P. Tun and A. Wingfield. One voice too many: Adult age differences in language processing with different types of distracting sounds. *The J. Gerontol. B. - Psychol.*, 54B(5):P317–P327, 1999. doi: 10.1093/geronb/54B.5.P317.

[172] P. Tun, G. O'Kane, and A. Wingfield. Distraction by competing speech in young and older adult listeners. *Psychol. Aging*, 17(3):453–467, 2002. doi: 10.1037/0882-7974.17.3.453.

[173] J. van Soest. Richtingshooren bij sinusvormige geluidstrillingen. *Physica*, 9:271–282, 1929.

[174] A. Vandierendonck, B. Liefooghe, and F. Verbruggen. Task switching: interplay of reconfiguration and interference control. *Psychol. Bull.*, 136(4): 601–626, 2010. ISSN 1939-1455. doi: 10.1037/a0019791.

[175] P. Verhaeghen and J. Cerella. Aging, executive control, and attention: a review of meta-analyses. *Neurosci. Biobehav. R.*, 26(7):849–857, 2002. ISSN 0149-7634. doi: 10.1016/S0149-7634(02)00071-4.

[176] F. Völk. Inter- and intra-individual variability in blocked auditory canal transfer functions of three circum-aural headphones. In *131st Audio Eng. Soc. Conv.*, New York, NY, US, 2011.

[177] H. Wallach. The role of head movements and vestibular and visual cues in sound localization. *J. Exp. Psychol.*, 27:339–368, 1940. doi: 10.1037/h0054629.

[178] C. Wasylyshyn, P. Verhaeghen, and M. Sliwinski. Aging and task switching: a meta-analysis. *Psychol. Aging*, 26(1):15, 2011. doi: 10.1037/a0020912.

[179] F. Wefers. *Partitioned convolution algorithms for real-time auralization: Dissertation*, volume 20 of *Aachener Beiträge zur Akustik*. Logos Verlag Berlin Gmbh, Berlin Germany, 2015. ISBN 978-3-8325-3943-6.

[180] E. Wenzel, M. Arruda, D. Kistler, and F. Wightman. Localization using nonindividualized head-related transfer functions. *J. Acoust. Soc. Am.*, 94 (1):111–123, 1993. doi: 10.1121/1.407089.

[181] F. Wightman and D. Kistler. Headphone simulation of free-field listening. I: Stimulus synthesis. *J. Acoust. Soc. Am.*, 85(2):858–867, 1989. doi: 10.1121/1.397557.

[182] F. Wightman and D. Kistler. Headphone simulation of free-field listening. II: Psychophysical validation. *J. Acoust. Soc. Am.*, 85(2):868–878, 1989. doi: 10.1121/1.397558.

[183] F. Wightman and D. Kistler. Resolution of front–back ambiguity in spatial hearing by listener and source movement. *J. Acoust. Soc. Am.*, 105(5): 2841–2853, 1999. doi: 10.1121/1.426899.

[184] N. Wood and N. Cowan. The cocktail party phenomenon revisited: Attention and memory in the classic selective listening procedure of cherry (1953). *J. Exp. Psychol. General*, 124(3):243–262, 1995. doi: 10.1037/0096-3445.124.3.243.

[185] W. Yost, R. Dye, and S. Sheft. A simulated "cocktail party" with up to three sound sources. *Atten. Percept. Psycho.*, 58(7):1026–1036, 1996. doi: 10.3758/bf03206830.

[186] P. Zahorik, F. Wightman, and D. Kistler. The fidelity of virtual auditory displays. *J. Acoust. Soc. Am.*, 99(4):2596(A), 1996. doi: 10.1121/1.415284.

Danksagungen

An dieser Stelle möchte ich allen danken, die mich in unterschiedlichsten Weisen während meiner Promotionszeit unterstützt haben.

In erster Linie gilt mein Dank Janina Fels, die mich in den Jahren meiner Promotion sehr freundschaftlich und immer hilfsbereit in fachlichen und privaten Fragen betreut und unterstützt hat.

Außerdem danke ich meinem Zweitberichter Iring Koch, der mir immer mit guten Ideen und Anregungen in zahlreichen gemeinsamen Publikationen zur Seite gestanden hat.

Die Forschung für die Dissertation wurde durch ein DFG-Gemeinschaftsprojekt mit der Psychologie finanziert. Auf Seiten der Psychologie hatte ich immer einen sehr netten und regen Austausch mit Vera Lawo und Julia Seibold.

Allen Kolleginnen und Kollegen des ITA gilt ebenfalls mein Dank für die schöne Atmosphäre und Zeit am ITA. Dabei sind aus Kollegen auch in einigen Fällen gute Freunde geworden. Neben unterhaltsamen Gesprächen in den Pausen mit Shaima'a Doma, Karin Loh und Michael Kohnen, wurde mit Martin Guski Einkaufen zu einem Event und mit Rob Opdam autotechnische Probleme zu einer netten Verabredung. Besonders aber werde ich meinen jahrelangen Bürokollegen Jan-Gerrit Richter vermissen.

Für die technische Umsetztung der Experimente danke ich Herrn Schlömer und Rolf Kaldenbach.

Danken möchte ich auch allen meinen Studierenden, insbesondere Britta Karnbach, Suliang Wang, Albina Kamil, Valeria Tarasova, Saskia Wepner und Dorothea Setzer, die mich in den Datenerhebungen und Probandenakquise unterstützt haben.

Zu guter Letzt möchte ich meiner Familie, insbesondere meinem Mann Simon, danken für ihre Unterstützung, ihr Verständnis und ihre Liebe. Es hat mir gut getan, dass meine Töchter, Margareta und Karolina, mich in schwierigen Phasen immer wieder auf den Boden der Tatsachen zurückgeholt haben und mir gezeigt haben was wirklich wichtig ist im Leben.

Curriculum Vitae

Personal Data

	Josefa Oberem
16.11.1986	born in Bonn-Beuel, Germany

Education

1993–1997	Primary School, "Erich-Kästner-Gemeinschaftsgrundschule", Bonn
1997–2006	Secondary School, "Otto-Kühne-Schule Pädagogium Godesberg", Bonn

Higher Education

2006–2010	Bachelors' Degree in Physics Friedrich-Wilhelms-Universität Bonn
2010–2012	Masters' Degree in Physics RWTH Aachen University Focus: Technical Acoustics and Particle Physics

Professional Experience

since Okt. 2012	Research Assistant, Institute of Technical Acoustics (ITA), RWTH Aachen (Parental time: Sept. 2014–Okt. 2015; Jun. 2017–Sept. 2018)

Aachen, Germany, March 22, 2020

Bisher erschienene Bände der Reihe

Aachener Beiträge zur Akustik

ISSN 1866-3052
ISSN 2512-6008 (seit Band 28)

1	Malte Kob	Physical Modeling of the Singing Voice ISBN 978-3-89722-997-6 40.50 EUR
2	Martin Klemenz	Die Geräuschqualität bei der Anfahrt elektrischer Schienenfahrzeuge ISBN 978-3-8325-0852-4 40.50 EUR
3	Rainer Thaden	Auralisation in Building Acoustics ISBN 978-3-8325-0895-1 40.50 EUR
4	Michael Makarski	Tools for the Professional Development of Horn Loudspeakers ISBN 978-3-8325-1280-4 40.50 EUR
5	Janina Fels	From Children to Adults: How Binaural Cues and Ear Canal Impedances Grow ISBN 978-3-8325-1855-4 40.50 EUR
6	Tobias Lentz	Binaural Technology for Virtual Reality ISBN 978-3-8325-1935-3 40.50 EUR
7	Christoph Kling	Investigations into damping in building acoustics by use of downscaled models ISBN 978-3-8325-1985-8 37.00 EUR
8	Joao Henrique Diniz Guimaraes	Modelling the dynamic interactions of rolling bearings ISBN 978-3-8325-2010-6 36.50 EUR
9	Andreas Franck	Finite-Elemente-Methoden, Lösungsalgorithmen und Werkzeuge für die akustische Simulationstechnik ISBN 978-3-8325-2313-8 35.50 EUR

10	Sebastian Fingerhuth	Tonalness and consonance of technical sounds ISBN 978-3-8325-2536-1 42.00 EUR
11	Dirk Schröder	Physically Based Real-Time Auralization of Interactive Virtual Environments ISBN 978-3-8325-2458-6 35.00 EUR
12	Marc Aretz	Combined Wave And Ray Based Room Acoustic Simulations Of Small Rooms ISBN 978-3-8325-3242-0 37.00 EUR
13	Bruno Sanches Masiero	Individualized Binaural Technology. Measurement, Equalization and Subjective Evaluation ISBN 978-3-8325-3274-1 36.50 EUR
14	Roman Scharrer	Acoustic Field Analysis in Small Microphone Arrays ISBN 978-3-8325-3453-0 35.00 EUR
15	Matthias Lievens	Structure-borne Sound Sources in Buildings ISBN 978-3-8325-3464-6 33.00 EUR
16	Pascal Dietrich	Uncertainties in Acoustical Transfer Functions. Modeling, Measurement and Derivation of Parameters for Airborne and Structure-borne Sound ISBN 978-3-8325-3551-3 37.00 EUR
17	Elena Shabalina	The Propagation of Low Frequency Sound through an Audience ISBN 978-3-8325-3608-4 37.50 EUR
18	Xun Wang	Model Based Signal Enhancement for Impulse Response Measurement ISBN 978-3-8325-3630-5 34.50 EUR
19	Stefan Feistel	Modeling the Radiation of Modern Sound Reinforcement Systems in High Resolution ISBN 978-3-8325-3710-4 37.00 EUR
20	Frank Wefers	Partitioned convolution algorithms for real-time auralization ISBN 978-3-8325-3943-6 44.50 EUR

21	Renzo Vitale	Perceptual Aspects Of Sound Scattering In Concert Halls ISBN 978-3-8325-3992-4 34.50 EUR
22	Martin Pollow	Directivity Patterns for Room Acoustical Measurements and Simulations ISBN 978-3-8325-4090-6 41.00 EUR
23	Markus Müller-Trapet	Measurement of Surface Reflection Properties. Concepts and Uncertainties ISBN 978-3-8325-4120-0 41.00 EUR
24	Martin Guski	Influences of external error sources on measurements of room acoustic parameters ISBN 978-3-8325-4146-0 46.00 EUR
25	Clemens Nau	Beamforming in modalen Schallfeldern von Fahrzeuginnenräumen ISBN 978-3-8325-4370-9 47.50 EUR
26	Samira Mohamady	Uncertainties of Transfer Path Analysis and Sound Design for Electrical Drives ISBN 978-3-8325-4431-7 45.00 EUR
27	Bernd Philippen	Transfer path analysis based on in-situ measurements for automotive applications ISBN 978-3-8325-4435-5 50.50 EUR
28	Ramona Bomhardt	Anthropometric Individualization of Head-Related Transfer Functions Analysis and Modeling ISBN 978-3-8325-4543-7 35.00 EUR
29	Fanyu Meng	Modeling of Moving Sound Sources Based on Array Measurements ISBN 978-3-8325-4759-2 44.00 EUR
30	Jan-Gerrit Richter	Fast Measurement of Individual Head-Related Transfer Functions ISBN 978-3-8325-4906-0 45.50 EUR
31	Rhoddy A. Viveros Muñoz	Speech perception in complex acoustic environments: Evaluating moving maskers using virtual acoustics ISBN 978-3-8325-4963-3 35.50 EUR
32	Spyros Brezas	Investigation on the dissemination of unit watt in airborne sound and applications ISBN 978-3-8325-4971-8 47.50 EUR

33	Josefa Oberem	Examining auditory selective attention: From dichotic towards realistic environments ISBN 978-3-8325-5101-8 43.00 EUR

Alle erschienenen Bücher können unter der angegebenen ISBN-Nummer direkt online (http://www.logos-verlag.de) oder per Fax (030 - 42 85 10 92) beim Logos Verlag Berlin bestellt werden.